The Lab Tutorial of Electrical Engineering and Electronics

电工电子学实验教程

主 编 杨永红 奚彩萍

江苏大学出版社
JIANGSU UNIVERSITY PRESS

镇 江

图书在版编目(CIP)数据

电工电子学实验教程 = The Lab Tutorial of Electrical Engineering and Electronics：英文 / 杨永红，奚彩萍主编. — 镇江：江苏大学出版社，2020.10
　　ISBN 978-7-5684-1433-3

Ⅰ. ①电… Ⅱ. ①杨… ②奚… Ⅲ. ①电工技术－实验－教材－英文②电子技术－实验－教材－英文 Ⅳ. ①TM－33②TN－33

中国版本图书馆 CIP 数据核字(2020)第 170623 号

电工电子学实验教程
The Lab Tutorial of Electrical Engineering and Electronics

编　　者/杨永红　奚彩萍
责任编辑/李菊萍
出版发行/江苏大学出版社
地　　址/江苏省镇江市梦溪园巷 30 号(邮编：212003)
电　　话/0511-84446464(传真)
网　　址/http://press.ujs.edu.cn
排　　版/镇江市江东印刷有限责任公司
印　　刷/江苏凤凰数码印务有限公司
开　　本/787 mm×1 092 mm　1/16
印　　张/11
字　　数/349 千字
版　　次/2020 年 10 月第 1 版　2020 年 10 月第 1 次印刷
书　　号/ISBN 978-7-5684-1433-3
定　　价/36.00 元

如有印装质量问题请与本社营销部联系(电话:0511-84440882)

Preface

The international education of science and engineering majors is currently developing rapidly. For science and engineering majors, electrical engineering and electronics including lab teaching is an essential professional course. At present, there are few publicly published english textbooks of electrical and electronic lab tutorials. We hope the textbook can promote international education and bilingual teaching in China.

The textbook is divided into 5 chapters. In Chapter 1, the course purpose and requirement are briefly introduced. The lab safe electricity notices are briefly explained. In Chapter 2, basic electronic components are comprehensively summarized, including resistors, capacitors, inductors, transistors, relays, field effect transistors, and thyristor. The structure and principle of some commonly used instruments are systematically outlined, such as oscilloscope, function generator, digital multimeter, electrodynamometer wattmeter, spectrum analyzer, and AC voltmeter. The measurement, significant figures, and measurement errors are universally demonstrated. The principles of AD and DA converter are mainly elaborated. Chapter 3 is the electrical engineering lab parts, including verification of circuit theorems, the measurement of 3-phase AC circuits, and the operation of induction motors and so on. Chapter 4 is electronics lab parts, including amplifier circuits, combinational logic circuit design, rectified DC power supply, and 555 timer applications. In Chapter 5, NI ELVISmx workstation and instrument launcher are briefly introduced. The interface with MATLAB is also elaborated.

The publication of the textbook is funded by Jiangsu University of Science and Technology. The authors are grateful to all these experts who have provided the treasured advices. Due to the limited level of authorship, there are inevitable faults and omissions in the textbook. We hope readers to point out and correct them.

Authors
9th Dec, 2020

Contents

Chapter 1　Introduction /001

　　1.1　Purpose and significance of lab course /001

　　1.2　Request of lab course /002

　　1.3　Laboratory safe electricity notice /002

Chapter 2　Electronic Components, Instruments, and Measurement Basis /004

　　2.1　Electronic components /004

　　　　2.1.1　Resistors /004

　　　　2.1.2　Capacitors /006

　　　　2.1.3　Inductors /008

　　　　2.1.4　Diodes /010

　　　　2.1.5　Bipolar junction transistors /011

　　　　2.1.6　Relays /013

　　　　2.1.7　Field effect transistors /014

　　　　2.1.8　Thyristors /016

　　2.2　Basic instruments /018

　　　　2.2.1　Oscilloscopes /018

　　　　2.2.2　Function generators /021

　　　　2.2.3　Digital multimeters /022

　　　　2.2.4　Electrodynamo watt meters /024

　　　　2.2.5　Spectrum analyzers /025

　　　　2.2.6　AC voltmeters /028

2.3　Measurement basis /030

    2.3.1　Significant figures /030

    2.3.2　Measurement /031

    2.3.3　Measurement error types /032

    2.3.4　AD converters /034

    2.3.5　DA converters /037

Chapter 3　Electrical Engineering Lab Parts /040

3.1　Voltage-current characteristic /040

3.2　Kirchhoff's law and superposition principle /043

3.3　Thevenin's theorem /046

3.4　Impedance and power factor improvement /049

3.5　Series RLC resonance circuits /053

3.6　Transient response in a first-order RC circuit /056

3.7　Measure voltage and current in 3-phase AC circuits /059

3.8　Power measurement in 3-phase AC circuits /063

3.9　Single-phase voltage transformers /067

3.10　Start 3-phase induction motors /071

3.11　Jog and continuous control of motors /075

3.12　Forward and reverse rotation of motors /079

3.13　Test dependent sources /082

3.14　The response of second order circuits /088

3.15　Negative impedance converters /091

3.16　PLC program for star-delta motor starter /094

Chapter 4　Electronics Lab Parts /097

4.1　Common emitter amplifier circuits /097

4.2　Long-tail pairs differential amplifier circuits /101

4.3　Negative feedback amplifier circuits /104

4.4　Integrated operational amplifier circuits /107

4.5　Applications of integrated operational amplifiers /112

4.6　Rectified DC power supplies /115

4.7   Test logic function of basic gates circuits /118

4.8   Test combinational logic circuits /122

4.9   Design combinational logic circuits /126

4.10 Test logic functions of flip flops /128

4.11 Sequential logic circuits /133

4.12 Digital counter, decoder and display circuits /136

4.13 555 timer applications /140

4.14 Waveform generation circuits /144

Chapter 5   NI ELVISmx Introduction /146

5.1   NI ELVISmx workstation /146

   5.1.1   Workstation connectors and controls /146

   5.1.2   Prototyping board /148

5.2   NI ELVISmx instrument launcher /148

5.3   Interface with MATLAB /155

   5.3.1   NI ELVISmx with data acquisition toolbox /155

   5.3.2   Lab: control stepper motor using digital outputs /157

Appendix /161

Appendix A   Electrical and electronic units /161

Appendix B   Metric prefix /162

Appendix C   Abbreviations in electrical engineering and electronics /163

Appendix D   NI ELVIS resource conflicts /164

Appendix E   NI ELVIS signal description /165

Reference /167

Chapter 1

Introduction

1.1 Purpose and significance of lab course

The electrical engineering and electronics is a basic major course of science and engineering majors, including electric circuits, analog electronics and digital electronics. In electric circuits parts, there are basic concepts and theorems of direct current (DC) circuits, sinusoidal alternate current (AC) circuit, 3-phase AC circuit analysis, transformers, and 3-phase AC induction motors. In analog electronics parts, there are semiconductor components, basic amplifier circuits, integrated operational amplifier (op amp) circuits, frequency response of amplifier circuits, negative feedback amplifier, waveform generation and signal conversion, power amplifier circuits, and DC power supplies. In digital electronics parts, there are logic algebra, combinational logic circuit, digital to analog converter and sequential logic circuit.

In lab course, students can verify the abstract and incomprehensible theories such as Thevenin theorem and superposition principle, analyze and generalize the characteristics of electrical components such as resistors, inductors, transistors and capacitors. It is beneficial to enhance students active learning and thinking, and to improve teaching standards. As a result, the lab course is an essential teaching link and bridges between theory and practice. Upon completion of the electrical engineering and electronics laboratory course, students can develop the following capabilities:

(1) The ability to properly select an AC / DC voltage source.

(2) The ability to analyze the measured data and error source based on circuit concepts and theorems.

(3) The ability to design analog circuits with certain functions based on resistors, capacitors, inductors, diodes, and transistors.

(4) The ability to design digital circuits with certain functions based on combinational logic circuits, flip flops, and 555 timer.

(5) The ability to use common electronics instruments refers to the knowledge and operating of function generators, digital multimeters, and oscilloscopes.

1.2 Request of lab course

Each lab session lasts 90 min. At the beginning of the lab, the instructor will give a brief introduction with demos. In order to proceed smoothly in the lab lesson, students should do the following works:

(1) Pre-lab: Students should carefully review the material in the textbook, understand the lab principle and procedure, and make a simple lab preview report including a lab scheme. There will be a short quiz at the first 5 minutes of the lab. The score of pre-lab parts is worth 30% of the overall lab score.

(2) In-lab: Each lab is done in groups and each group is divided into two students. Teamwork approach encourages interaction and facilitates debugging and data collection. Each student has his/her own lab notebook for recording the pre-lab and measurement data. Copying data from other groups or altered measurement data will result in a zero grade. The score of in-lab parts is worth 40% of the overall lab score.

(3) Pro-lab: Keep the workbench clean and put everything back its proper place after the lab is over. Finish lab report after class and submit the lab report to lab instructor on time. The score of pro-lab parts is worth 30% of the overall lab score.

In addition, no foods are allowed in the lab for any reason. Don't be late for 10 minutes. The circuit components and instruments are available in the lab, such as oscilloscope, function generator, resistors, capacitors, transistors, integrated chips (IC), wires, cables, and connectors. If incorrectly used, these may be damaged. Think twice before connecting the meter. In particular, check the switch position and ensure that the test leads are connected to the correct input on the meter.

1.3 Laboratory safe electricity notice

In the laboratory, please obey the following general rules and precautions which are benefit for both experimenters and those around them.

(1) Students should carry notebook, record the data, take picture, and draw the curves on the graph sheet.

(2) Check circuit power supply voltage source type DC/AC and frequency parameter before starting the lab.

(3) Students should take care of the valuable laboratory instruments.

(4) Before plugging into an outlet, be sure that the power switch is turned off.

(5) Before unplugging the power from the outlet, be sure that the power button of the instrument or equipment is on "OFF" position.

(6) All conducting surfaces intended to be at ground potential should be connected together.

(7) Students should ensure that all switches are in the "OFF" position and remove the connections before leaving the laboratory.

(8) Remove conductive chains, finger rings and wrist watches.

(9) Do not use metallic pencils, metal or metal edge rulers.

(10) Do not wear long loose ties, scarves, or other loose clothing around machines.

(11) Unplug the power cord instead of the cable.

(12) Keep liquids and chemicals away from instruments and circuits.

(13) Make measurements with one hand at a time. Do not touch any part of the live circuit with bare hands.

(14) Keep the body or any part of the body away from the circuit.

(15) Where interconnecting wires and cables are involved, they should be arranged so as not to trip over people.

(16) Report any equipment damage, hazards, and potential hazards to the laboratory instructor.

In addition, the following unsafe items should also be cautioned or checked in the lab. If any of the following happens, report it to the lab instructor immediately.

(1) Is the cord's insulation cracked or damaged, exposing the internal wiring?

(2) Are the plug's prongs broken or missing?

(3) Is the plug or outlet blackened by arcing?

(4) Are any protective parts broken, cracked or missing?

(5) Does the equipment and cord overheat?

(6) Does the equipment spark when switching on?

(7) Do you feel a light shock during the lab work?

(8) Is liquid spilled around the equipment?

Chapter 2

Electronic Components, Instruments, and Measurement Basis

2.1 Electronic components

2.1.1 Resistors

Resistors are passive electric components that reduce the flow of electrical current in a circuit. As shown in Figure 2-1-1, there are fixed resistors, variable resistors, thermistors, varistors, photoresistors, and magneto-resistors. The application fields of resistors include generating heat, delimit current or create a voltage drop.

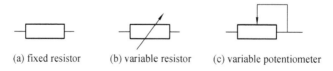

(a) fixed resistor (b) variable resistor (c) variable potentiometer

Figure 2-1-1 The circuit symbols of resistors

Resistivity is a measure of the resistance of a given size of a specific material to electrical conduction. Resistivity of common materials at temperature 23℃ is listed in Table 2-1-1. Depending on the length, cross sectional area, and material resistivity, the resistance is expressed by

$$R = \rho \frac{l}{A} \tag{2-1-1}$$

where l is the length of the material, ρ is resistivity, and A is the cross-sectional area of the material. The unit of resistance is expressed in Ohms (Ω).

Table 2-1-1 Resistivity of common materials at temperature 23℃

material	resistivity/($\Omega \cdot m$)	usage
gold	2.45×10^{-8}	conductor
silver	1.59×10^{-8}	conductor
copper	1.68×10^{-8}	conductor
aluminum	2.65×10^{-8}	conductor
iron	9.71×10^{-8}	conductor
germanium	0.47	semiconductor
silicon	640	semiconductor
mica	5×10^{11}	insulator
glass	10^{12}	insulator

As shown in Figure 2-1-2, the use of color bands on the body of a resistor is the most common system for indicating the value of a resistor. Most resistors have four bands, but there are also resistors with three, five, and six bands. Color-coding is standardized by the electronic industries association. Resistor color code is listed in Table 2-1-2.

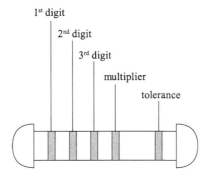

Figure 2-1-2 The resistor color code

Table 2-1-2 Resistor color code

color	1st digit	2nd digit	3rd digit	multiplier	tolerance
black	0	0	0	1	—
brown	1	1	1	10	±1%
red	2	2	2	10^2	±2%
orange	3	3	3	10^3	±3%
yellow	4	4	4	10^4	±4%
green	5	5	5	10^5	±0.5%
blue	6	6	6	10^6	±0.25%
violet	7	7	7	10^7	±0.1%
grey	8	8	8	10^8	±0.05%
white	9	9	9	10^9	—
gold	—	—	—	0.1	±5%
silver	—	—	—	0.01	±10%

Before selecting a resistor in a given circuit, you need to understand its specifications. Some basic specifications of resistors are briefly given as follows.

(1) *Resistance value*: This specification defines the rated resistance value is expressed by color code bands.

(2) *Tolerance*: This specification defines deviation between quoted and actual value of a resistor, which is expressed by color code bands.

(3) *Power rating*: This specification indicates the maximum power that the resistor can dissipate.

(4) *Maximum voltage*: The maximum value of DC voltage or maximum root mean square (rms) of AC voltage can be applied continuously to resistors at the rated ambient temperature.

(5) *Temperature coefficient*: The specification indicates how much resistance value changes with temperature, which is usually expressed in ppm/℃.

2.1.2 Capacitors

The capacitors are formed with two parallel metal electrodes separated by a non-conductive material. As shown in Figure 2-1-3, there are fixed capacitors, variable capacitors and electrolytic capacitor. In contrast to inductor, a capacitor blocks any DC component present in an AC signal. The application fields of capacitors include storing charges, conducting alternate current, and blocking different voltages level of DC source in electrical circuit.

(a) fixed capacitor   (b) variable capacitor   (c) electrolytic capacitor

Figure 2-1-3   The circuit symbols of capacitors

Depending on the plates size and a non-conductive material between the plates, the capacitance of a parallel plate capacitor is expressed by

$$C = \varepsilon_0 \varepsilon_r \frac{A}{d} \tag{2-1-2}$$

where d is the distance between two plates, A is the area of a plate, ε_0 is the dielectric constant of free space, and ε_r is the dielectric constant of non-conductive material between the plates. The basic unit of capacitance is the Farad (F).

Dielectric constants of commonly used materials are listed in Table 2-1-3. The capacitor materials include electrolytic, ceramic, film, tantalum, and silver mica. Each type of capacitor has its own advantages and disadvantages.

Table 2-1-3   Dielectric constants of commonly used materials

dielectric material	dielectric constant
air	1.0006
aluminum phosphate	6.0
paper (dry)	2.0
mica	7.0
tantalum oxide	11.6

capacitance values are divided into text marking code and color code. As shown in Figure 2-1-4, capacitors are labeled with alphabets and digits. The first two digits are the significant digits of the capacitance value, the third digit is the multiplier, and the fourth symbol is the tolerance. When no units are given, this value is in pico Farad (pF). Thus, the capacitance value of the capacitor in Figure 2-1-4 is 10×10^4 pF (100 nF), whose tolerance symbol "K" means ±10% between the rated value and actual value. Marking code of tolerance capacitor is listed in Table 2-1-4.

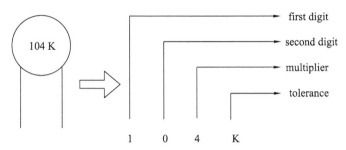

Figure 2-1-4 The text marking code of a capacitor

Table 2-1-4 Tolerance capacitor marking code

symbol	tolerance	symbol	tolerance
A	±0.05pF	K	±10%
B	±0.1pF	L	±15%
C	±0.25pF	M	±20%
D	±0.5nH	N	±30%
E	±0.5%	P	−0%, +100%
F	±1%	S	−20%, +50%
G	±2%	W	−0%, +200%
H	±3%	X	−20%, +40%
J	±5%	Z	−20%, +80%

Before selecting a capacitor in a given circuit, you need to understand its specification. Some basic specifications of capacitors are given as follows.

(1) *Capacitance value*: This specification defines the rated capacitance value expressed by color or alphanumeric codes.

(2) *Tolerance*: This specification defines deviation between quoted and actual value of a capacitor, which is expressed by color or alphabet codes.

(3) *Dielectric*: This specification refers to the dielectric characteristics of non-conductive materials. The dielectric is one of the key items that governs many of the capacitor characteristics.

(4) *Maximum voltage*: The maximum rms of AC voltage can be applied continuously to

capacitors at the rated ambient temperature.

(5) *Equivalent series resistance*: This specification refers to the impedance of the capacitor to alternating current, including the resistance of the dielectric material and the capacitor plates.

(6) *Leakage current*: The dielectric materials used in capacitors are not ideal insulators. A small direct current can flow or leak through the dielectric material for various reasons.

(7) *Ripple current*: This specification indicates the rms value of AC flowing through the capacitor.

(8) *Self-inductance*: There is self-inductance due to the impure capacitors.

(9) *Self-resonant frequency*: The self-resonant frequency of a capacitor is generated by a resonant circuit established between the equivalent series inductance and the capacitance of the capacitor.

2.1.3 Inductors

An inductor is just a coil of wire around some kind of core. The core could be just air or a magnet, iron. As shown in Figure 2-1-5, there are fixed inductor, variable inductor, and iron core inductor. In contract to a capacitor, an inductor blocks any AC component present in a DC signal. Inductors are used with capacitors in various wireless communications applications. An inductor connected in series or parallel with a capacitor can provide discrimination against unwanted signals. Large inductors are used in the power supplies of electronic equipment of all types, including computers and their peripherals. In these systems, the inductors help to smooth out the rectified utility AC, providing pure, battery-like direct current.

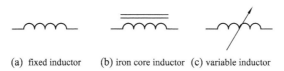

(a) fixed inductor   (b) iron core inductor   (c) variable inductor

Figure 2-1-5 The circuit symbols of inductors

Inductor is based on the theory of electromagnetic induction. It means any varying electric current, flowing in a conductor, produces a magnetic field around that, which is perpendicular to the current. Also, any varying magnetic field, produces current in the conductor present in that field, whereas the current is perpendicular to the magnetic field. The inductance is expressed by

$$L = \mu_0 \mu_r \frac{N^2 A}{l} \qquad (2\text{-}1\text{-}3)$$

where l is the average length of coils, A is the area of a coil, N is the number of turns the coil, μ_0 is the permeability of free space, and μ_r is the relative permeability. The basic unit of inductance is the Henry (H).

Inductance values are classified into text marking code and color code. As shown in Figure 2-1-6, an inductor is labeled with alphabets and digits. The first two digits are the significant digits of the inductance value, the third digit is the multiplier, and the fourth symbol is the tolerance. When no units are given, this value is in micro Henry (μH). Thus, the inductance value of the inductor in Figure 2-1-6 is 24×10^3 μH (24 mH), whose tolerance symbol "K" means ±10% between the rated value and actual value. Marking code of tolerance inductor is listed in Table 2-1-5.

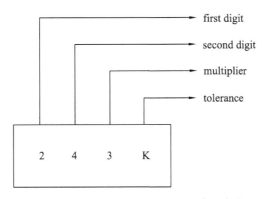

Figure 2-1-6 The text marking code of an inductor

Table 2-1-5 Marking code of tolerance inductor

symbol	tolerance	symbol	tolerance
B	±0.15nH	J	±5%
C	±0.2nH	K	±10%
S	±0.3nH	L	±15%
D	±0.5nH	M	±20%
F	±1%	V	±25%
G	±2%	N	±30%
H	±3%		

Before selecting a inductor in a given circuit, you need to understand its specifications. Some basic specifications of inductors are given as follows.

(1) *Inductance value*: This specification defines the rated inductance value expressed by color or alphanumeric codes.

(2) *Tolerance*: This specification defines deviation between quoted and actual value of an inductor, which is expressed by color or alphanumeric codes.

(3) *Permeability*: Higher permeability core materials cause inductors to provide higher levels of inductance. The material, shape, size, and geometry of the core affect the overall effective magnetic permeability.

(4) *DC resistance*: Because inductors are usually made from very thin wires, DC resistance can sometimes be very large. In most circuit simulations, the DC resistance can be thought of as being in series with a pure indicator, even though it is actually distributed throughout the inductor.

(5) *Maximum voltage*: This specification indicates the maximum voltage can be applied continuously to inductors at the rated ambient temperature.

(6) *Maximum DC current*: This specification refers to the maximum level of continuous direct current that can be passed through an inductor with no damage.

(7) *Winding self-capacitance*: This specification indicates that the wire has a small but appreciable level of capacitance.

2.1.4 Diodes

The semiconductor diode is widely used in many areas of electronics, such as power rectification, signal detection, light generation, and laser light generation. As shown in Figure 2-1-7, there are light emitting diodes, photodiode, PN junction, schottky diodes, tunnel diode and zener diode (voltage reference diode). Whatever the type of diode, the basic idea of the diode is a PN junction made from the semiconductor materials such as silicon, germanium, and gallium arsenide. The P layer has an abundance of holes (positive), and the N layer has an abundance of electrons (negative).

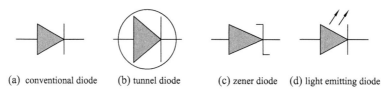

(a) conventional diode   (b) tunnel diode   (c) zener diode   (d) light emitting diode

Figure 2-1-7 **The circuit symbols of diodes**

Voltage-current characteristic of diode is illustrated in Figure 2-1-8. There are reverse region and forward region for a diode. Under forward bias, a p-type semiconductor material is connected to the positive electrode of the power supply, and a n-type semiconductor material is connected to the negative electrode of the power supply. When the voltage is increased by changing the resistance value, the circuit curve increases very slowly, and the curve becomes non-linear. Because the applied external voltage is used to cross the potential barrier of the PN junction diode, the current slowly rises in forward bias. However, when the potential barrier is completely eliminated and the external voltage applied to the junction increases, the PN junction behaves like a normal diode, and the circuit current increases sharply.

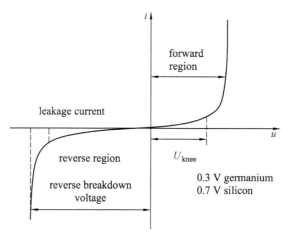

Figure 2-1-8 **The voltage-current characteristic of a diode**

Under reverse bias, the p-type material is connected to the negative pole of the power supply, and the n-type material is connected to the positive pole of the power supply. The resistance of the diode becomes very high, and virtually no current flows through the diode. This current is called reverse current. In this state, the resistance of the barrier increases because of breakdown at the junction. If the reverse bias current increases, the junction will be permanently damaged.

Before selecting a diode in a given circuit, you need to understand its specifications. Some basic specifications of diodes are given as follows.

(1) *Forward voltage drop*: This specification indicates that the voltage across a PN junction diode arise for the normal resistive losses and the depletion layer to be overcome and allows current to flow.

(2) *Peak inverse voltage*: This specification indicates that the characteristic of the diode is the maximum reverse voltage which the diode can withstand. Do not exceed this voltage, or the device may fail.

(3) *Maximum forward current*: This specification defines the diode maximum current which cannot be exceeded.

(4) *Leakage current*: For a true PN junction diode, due to minority carriers in the semiconductor, very little current flows in the opposite direction.

(5) *Reverse breakdown voltage*: This specification means that the breakdown voltage is the minimum reverse voltage which makes the diode conduct appreciably in reverse.

(6) *Junction capacitance*: This specification means that the capacitance is associated with the charge variation in the depletion layer, which is dependent upon the reverse voltage.

(7) *Junction operating temperature*: This specification defines that diodes have a maximum operating temperature.

2.1.5 Bipolar junction transistors

The transistor was invented by three scientists at the Bell Laboratories in 1947, which rapidly replaced the vacuum tube as an electronic signal regulator. Transistors are the active components of integrated circuits. There are two types of transistors in present: bipolar junction transistor (BJT) and field effect transistors (FET). The application fields of a common transistor include signal amplifiers, digital and analog switches, power regulators and equipment controllers.

As shown in Figure 2-1-9, there are NPN type and PNP type BJT with three terminals for connection to external circuits. Three leads on a bipolar transistor are named by collector, emitter and base.

(1) *Collector*: This lead is attached to the largest of the semiconductor regions. Current flows through the collector to the emitter as controlled by the base.

(2) *Emitter*: This lead is attached to the second largest of the semiconductor regions. When the base voltage allows, current flows through the collector to the emitter.

(3) *Base*: This lead is attached to the middle semiconductor region. This region serves as the gatekeeper that determines how much current is allowed to flow through the collector-emitter circuit.

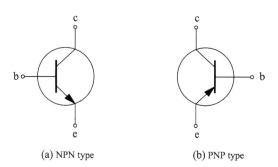

(a) NPN type   (b) PNP type

Figure 2-1-9 The circuit symbols of bipolar junction transistors

In addition, collector-emitter current and base-emitter current are essential in a transistor. The functions are briefly given as follows.

(1) *Collector-emitter*: The main current that flows through the transistor. Voltage placed across the collector and emitter is often referred to as U_{ce}, and current flowing through the collector-emitter path is called I_{ce}.

(2) *Base-emitter*: The current path that controls the flow of current through the collector-emitter path. Voltage across the base-emitter path is referred to as U_{BE} and is also sometimes called bias voltage. Current through the base-emitter path is called I_{BE}.

As shown in Figure 2-1-10, there are active region, saturation region, and cut-off region. The region between cut off and saturation is known as active region. In the active region, base-emitter junction remains forward biased and collector-base junction remains reverse biased. Therefore, the transistor will function normally in this region. A quiescent point is a point where the device characteristics and the load line intersect. The correct quiescent point of the transistor amplifier can reduce distortion of the output signal, such as the saturation distortion and cut off distortion.

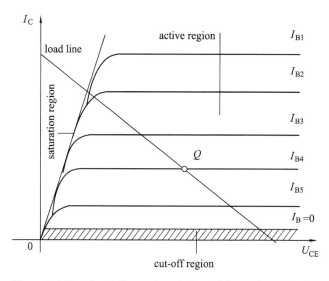

Figure 2-1-10 Load line and variation of the quiescent point

Before selecting a bipolar junction transistor in a given circuit, you need to understand its specifications. Some basic specifications of bipolar junction transistor are briefly given as follows.

(1) *Maximum collector dissipation* P_D: The total power can be dissipated by the device.

(2) *Current gain* β (h_{FE}): The ratio of the base current to the collector current. This is the amplifying ability of the transistor.

(3) *Collector cut off current* I_{CEO}: Leakage current from collector to emitter with base open circuit.

(4) *Collector saturation voltage* $V_{CE}(sat)$: The saturation voltage across the collector-emitter path.

(5) *Collector to emitter cut off voltage* V_{CEO}: The maximum voltage across the collector and the emitter.

(6) *Maximum collector current* $I_C(max)$: The maximum current flow through the collector-emitter path.

(7) *Maximum collector to emitter voltage* $V_{CE}(max)$: The maximum voltage across the collector and emitter.

(8) *Emitter-base voltage* V_{EBO}: The maximum voltage across the emitter and the base collector and emitter.

(9) *Collector to emitter breakdown voltage* BV_{CBO}: The breakdown voltage across the collector and emitter.

(10) *Base emitter saturation voltage* $V_{BE}(sat)$: The saturation voltage across the base and emitter.

(11) *Collector base cut off current* I_{CBO}: The cut off current flow through the collector-base path.

2.1.6 Relays

A relay is a simple electro-mechanical switch which closes or opens the circuit through electrical and mechanical operation. The main operation of the relay occurs where the low power signal can only be used to control the circuit. According to the operating principle and structural features, relays are classified into electromagnetic relays, overload relays, mechanical relays, reed relays, power varied relays, and solid-state relays. In order to against sudden current spikes, overload relay is used to safeguard motors from overloads and power failures. The thermal overload relay consists of a bimetallic strip which is to connect with a contact. When the excess current causes rising temperatures, the contact between the strip and the contact can de-energize the control circuit and switch off the motor power.

As shown in Figure 2-1-11, a relay consists of wounding a copper coil on a metal core, movable armature, and a set of electrical contacts which are named as normally open (NO), normally closed (NC) and common (COM) contacts. a NO contact closes the circuit when the relay is activated and disconnects the circuit when the relay is inactive. In contract to the NO contact, a NC contact disconnects the circuit when the relay is activated and connects the circuit when the relay is deactivated.

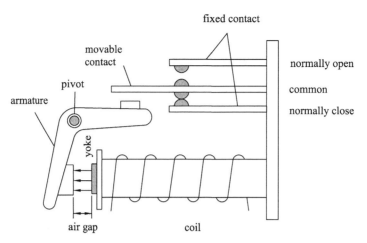

Figure 2-1-11 The block diagram of an electromagnet

Before selecting a relay, you need to understand its specifications. Some basic relay specifications are given as follows.

(1) *Coil resistance*: This is one of key parameters. When the relay is activated, the coil resistance determines the current flowing through it. Because the resistance of a coil varies with temperature, it is normally quoted at a temperature of 25 ℃.

(2) *Coil voltage*: This specification means a voltage across the coil required in order for the relay to switch correctly. It is normally standard voltages, such as 3, 5, 12, and 24 V.

(3) *Maximum carry current*: This specification means the relay can withstand the maximum current after switching on. It should not be exceeded, otherwise contacts will be damaged and shorten the life expectancy.

(4) *Maximum switch current*: This specification defines the relay can withstand the maximum switching current. In addition, the switching current is always less than the carry current.

(5) *Maximum switch voltage*: This specification indicates the relay can withstand the maximum switching voltage. In addition, the switching voltage is always less than the carry voltage.

(6) *Operate time*: It is taken the time from the application of the coil voltage to the settled state of relay contacts, which is determined by the rate of rise of the magnetic field. Moreover, the magnetic field is determined by the coil inductance and the inertia of the reed switch blades.

(7) *Release time*: It is taken the time for a relay contact to open after the operating coil has been de-energized.

2.1.7 Field effect transistors

In contrast to BJT, the field effect transistor (FET) uses an electric field to vary the depletion width during reverse biasing of the junction. The conduction mechanism in FETs is only due to the majority of charge carriers. Therefore, FETs are called unipolar devices. The application fields of FETs include analog switch, amplifier, and phase shift oscillator. As shown in Figure 2-1-12, FET consists of a semiconductor channel with three electrodes called drain, gate and source.

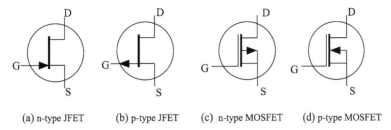

Figure 2-1-12 The circuit symbol of FETs

(1) *Channel*: This is the region through which majority charge carriers flow from source to drain. It can be either n-type or p-type.

(2) *Source*: It is the terminal that leads most of the charge carriers into the FET.

(3) *Drain*: It is the collection terminal of majority charge and can improves the conduction process.

(4) *Gate*: The gate terminal is a high impurity region and is formed by diffusion of one semiconductor with another semiconductor. It can control the carrier flow from the source to the drain.

The FETs are mainly divided into two types: junction field effect transistor (JFET) and insulated gate field effect transistor. The structure of JFET is illustrated in Figure 2-1-13. The current flowing through the drain and source is dependable on the voltage applied to the gate terminal. The gate voltage of n-type channel JFET is negative, and the gate voltage of p-type channel JFET is positive. When the junction of JFET is reversed biased, the conduction is established by variation of depletion width. As a result, JFET only works in the depletion mode.

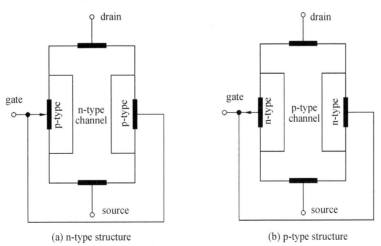

Figure 2-1-13 The structure of a JFET

Metal oxide semi-conductor FET (MOSFET) is one of the typical insulated gate FET in which the oxide layer plays a vital role. The MOSFET works in depletion mode and enhancement mode. As shown in Figure 2-1-14, there is a physical channel in depletion MOSFET, but not in enhancement MOSFET. Compared to the JFET, MOSFET has much higher input impedance due to small leakage current. As a result, it can be easily used in case of high current applications.

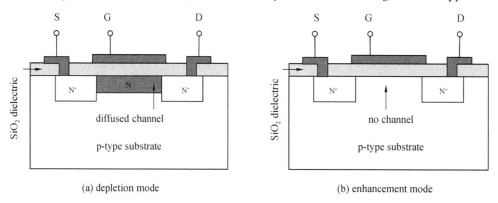

Figure 2-1-14 The structure of a n channel MOSFET

Before selecting a FET, you need to understand its specifications. Some basic FET specifications are briefly given as follows.

(1) *Gate source voltage* V_{GS}: This specification refers to a rating of the maximum voltage that can be withstood between the gate and source terminals.

(2) *Drain-Source Voltage* V_{DSS}: This specification indicates the maximum rating of the drain-source voltage can be tolerated without causing avalanche breakdown.

(3) *Input capacitance* C_{iss}: This specification defines the capacitance that is measured between the gate and source terminals with the drain shorted to the source for AC signals.

(4) *Drain-source on resistance* $R_{DS(on)}$: This specification defines the resistance across the channel between the drain and source when the FET is on.

(5) *Drain current at zero gate voltage* I_{DSS}: This specification defines the drain current for zero bias is the maximum current which flows through FET when the FET is on.

(6) *Gate source cut-off voltage* $V_{GS(off)}$: This specification defines the voltage which is needed at the gate-source region in order to switch off.

(7) *Power dissipation* P_{tot}: This specification defines the maximum continuous power that the device can dissipate.

(8) *Threshold voltage* $V_{GS(TH)}$: This specification refers to the minimum gate voltage that can provide a conductive channel between the gate and source.

2.1.8 Thyristors

In contrast to the junction diode and the bipolar transistor, the thyristor or silicon-controlled rectifier (SCR) is a four-layer (P-N-P-N) semiconductor device that contains three PN junctions in series (J_1, J_2, and J_3) as shown in Figure 2-1-15. It has three terminals called anode, cathode and gate. The main application fields of thyristor include light controls, electric power controls and motor speed controls.

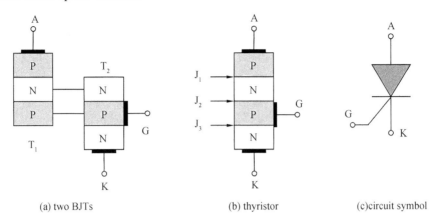

(a) two BJTs　　　　(b) thyristor　　　　(c)circuit symbol

Figure 2-1-15　The structure and circuit symbol of a thyristor

The voltage-current characteristic of a thyristor is illustrated in Figure 2-1-16. There are reverse blocking region and forward blocking region, and forward conduction region. Once the thyristor is "turned on", the current will be flowed in the forward direction. The gate signal will lose all control effect because of the regenerative latching effect of the two internal transistors. Applying any gate signal or pulse after regeneration starts will be completely invalid because the thyristor is already on and fully on. Like the diode, the thyristor is a unidirectional device. However, thyristors can only operate in the switching mode and cannot be used for amplification.

In order to switch off the thyristor, the anode current must be reduced below this minimum level and hold on long enough for the internally latched PN-junctions to recover their blocked state.

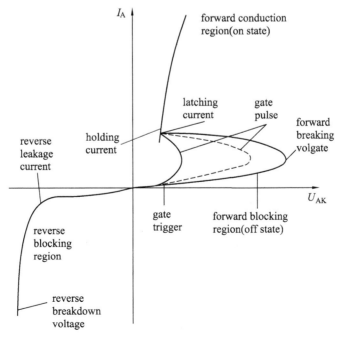

Figure 2-1-16　The voltage-current characteristic of a thyristor

Before selecting a thyristor, you need to understand its specifications. Some basic thyristor specifications are briefly given as follows.

(1) *Peak inverse voltage*: This specification defines the maximum voltage across the SCR can safely be withstood in its OFF state. Under no circumstances should the applied voltage be exceeded.

(2) *ON state voltage*: This specification indicates the voltage across the SCR appears during its on-state.

(3) *Minimum gate current*: This specification indicates the minimum current required at the gate for triggering the SCR.

(4) *Maximum gate current*: This specification indicates the maximum current applied to device safely.

(5) *Latching current*: This specification refers to the minimum current required to latch the device from its OFF state to its ON state.

(6) *Holding current*: This specification refers to the minimum current required to hold the SCR conduct.

(7) *Turn ON time*: It is taken the time for the device to change from its OFF state to its ON state.

(8) *Turn OFF time*: After applying reverse voltage, the device takes a finite time to get switched OFF state.

(9) *Gate power loss*: It is caused by gate current between the gate and the main terminals.

(10) *Maximum surge-ON state current*: This specification refers to the maximum allowable peak value of a sinusoidal half cycle period at a frequency of 50 Hz.

2.2　Basic instruments

2.2.1　Oscilloscopes

The main purpose of an oscilloscope is to display electrical signals over time. Oscilloscopes generate a two-dimensional plot where the x-axis is time and the y-axis is voltage. Oscilloscopes are divided into analogue oscilloscope, analogue storage oscilloscope, digital domain oscilloscope, digital phosphor oscilloscope, digital sampling oscilloscope and mixed signal oscilloscope. They are essential instruments in the education fields of science and engineering discipline.

As shown in Figure 2-2-1, the analogue scope is divided into cathode ray tube (CRT), amplifier, vertical channel, time base generator and trigger circuit. A cathode ray tube consists of electron gun, deflection plates, fluorescent screen, and glass case. The main function of the electron gun is to generate the electron beam. The electron gun includes heater, cathode electrodes, control grid and acceleration anodes. The cathode electrode is heated by the heater and emits the electrons. The electrons pass through the control grid, are accelerated by pre-acceleration and acceleration anode, and are focused by focusing anode. The electron beam flows through the deflection plates and reaches the fluorescent screen.

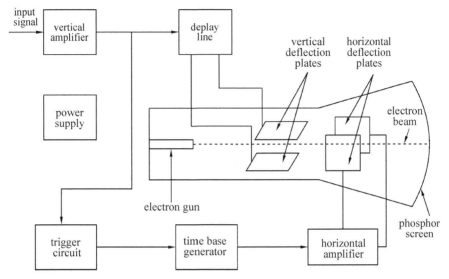

Figure 2-2-1 The block diagram of an analogue scope

The deflection system consists of two pairs of parallel plates referred to as the horizontal and vertical deflection plates. The input signal is connected to vertical deflection plates, and the time base is connected to horizontal deflection plates. As shown in Figure 2-2-2, the deflection distance of the electron beam on the screen in the Y direction is denoted by "Y". The deflection distance is expressed by

$$Y = \frac{l_y L}{2dU_a} U_y = \frac{1}{k} U_y \qquad (2\text{-}2\text{-}1)$$

where U_a is acceleration anode voltage, U_y is voltage between deflecting plates, d is distance between deflecting plate, l_y is the deflecting plate length, L is the distance between screen and the mid of the deflecting plates, k is the deflection coefficient. The Equation 2-2-1 shows the deflection distance of the electron is directly proportional to the deflecting voltage.

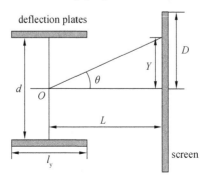

Figure 2-2-2 The deflection distance

Before using an oscilloscope, you need to understand its specifications. Some basic oscilloscope specifications are briefly given as follows.

(1) *Bandwidth*: Oscilloscopes are most commonly used to measure waveforms with a certain frequency. The scope's bandwidth specifies the frequency range. Often a five times rule is used as a rule of thumb.

(2) *Number of channels*: Oscilloscopes can read multiple signals at once and display them all on the screen at the same time. A two-channel range is commonly used.

(3) *Maximum input voltage*: Each electronic component has a high voltage limit. The oscilloscope should have a maximum input voltage. If the signal exceeds the maximum input voltage, the oscilloscope may be damaged.

(4) *Vertical sensitivity*: This specification defines the minimum and maximum values of the vertical voltage. The value is measured by volts per division.

(5) *Time base*: The time base usually indicates the sensitivity range on the horizontal time axis. The value is measured by seconds per division.

(6) *Resolution*: The resolution deflection is the accuracy of the input voltage. This value can be changed by adjusting the vertical scale.

(7) *Input impedance*: For high-frequency signals, each oscilloscope adds a certain impedance to the circuit it reads, called the input impedance.

(8) *Rise time*: This specification defines the fastest rise pulse it can measure. Oscilloscope rise time is closely related to bandwidth.

(9) *Sample rate*: This feature is unique to digital oscilloscopes and defines how many times a signal is read per second. For oscilloscopes with multiple channels, this value may decrease if multiple channels are used.

The oscilloscope is a basic instrument in labs. Students should be familiar with the usage of the oscilloscope. The operation steps of an oscilloscope are as follows.

(1) Switch on the power button. Wait for the CRT to warm up and begin to operate appeared.

(2) Adjust the brightness control button up until a horizontal line appears.

(3) Adjust the focus control button until a thin line appeared.

(4) Connect a probe into the CH_1 input and the probe tip into the CAL output.

(5) Adjust the time base control button until a square wave is appeared.

(6) Adjust the trimmer on the probe until the overshoot or undershoot disappear.

(7) Remove probe tip from the square wave output. Calibration is completed.

(8) It is now ready to use the oscilloscope to observe signal waveform.

(9) Plug the coaxial cable into the oscilloscope. Connect the coaxial cable (usually red color) to the signal, and the alligator clip (usually black color) to ground.

(10) Adjust the signal waveform on the screen by pressing or rotating the corresponding buttons.

(11) Measure the amplitude and frequency of signal by pressing or rotating the corresponding buttons.

2.2.2 Function generators

A signal generator is an electronic test instrument that generates either repeating or non-repeating waveforms. Signal generators are mostly used in designing, manufacturing and repairing electronic devices. The signal generator types are divided into function generators, RF and microwave signal generators, pitch generators, arbitrary waveform generators, digital pattern generators and frequency generators. In general, no device is suitable for all possible applications. As shown in Figure 2-2-3, the main structure of function generator consists of a RC integrator, constant current supply sources, a frequency control network, amplifiers, and voltage comparators.

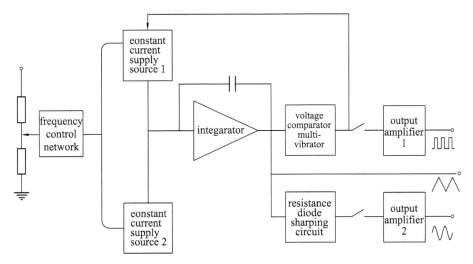

Figure 2-2-3 The block diagram of a function generator

The function generator is used to generate simple repetitive waveforms such as sine waves, sawtooth wave, square wave and triangular waveforms. In addition to select the waveforms, other control buttons on the function generator may include:

(1) *Frequency*: This control button alters the basic frequency where the waveform repeats.

(2) *Waveform type*: This control button enables different basic waveform types to be selected, such as sine wave, square wave and triangular wave.

(3) *DC offset*: This control button alters the average voltage of a signal relative to 0 V.

(4) *Duty cycle*: This control button changes the ratio of high voltage time to low voltage time in a square wave signal, and a triangular waveform with equal rise and fall times to a sawtooth.

Before using a function generator, you need to understand its specifications. Some basic specifications of a function generator are briefly given as follows.

(1) *Waveforms*: Function generators usually create sine, square, pulse, triangle, and sawtooth waves. In addition, some specifications should be considered, such as sine wave distortion, square wave symmetry, triangle wave linearity, square wave rise and fall times.

(2) *Output level*: The output level of commonly used signal will change continuously. When

creating TTL signal, its peak value may be adjusted between 0 to 12 V.

(3) *Output impedance*: The load that can be driven by the function generator is very important. The load is typically 50 Ω.

(4) *Bandwidth*: The bandwidth is limited by the output amplifier or the filters in the analog output circuit. This specification determines the maximum output frequency of the signal generator.

(5) *DC offset*: This allows the basic voltage level of the signal to vary within a given range, such as between −5 V and +5 V.

(6) *Frequency range*: The frequency range of the function generator is limited. The lower limit frequency is usually below 1 Hz. Generally, the lower limit can be met. The upper limit frequency can be changed in the frequency range of 1 MHz to 20 MHz or higher.

(7) *Frequency stability*: The frequency stability of a function generator can vary widely. For a function generator, a typical value may be about 0.1% per hour.

(8) *Phase lock capability*: This specification can allow the function generator to provide a more accurate and synchronized output.

2.2.3 Digital multimeters

As shown in Figure 2-2-4, a digital multimeter (DMM) can measure AC/DC voltage, AC/DC current and resistance. As a result, it is one of the most indispensable instruments in electrical engineerig and electronics lab. In Figure 2-2-4, the input terminal is connected to the mode selector knob. The other input terminal is a common terminal, which is often called a ground terminal or a negative terminal. The mode selector switch has five positions. Each position of the switch is labeled with its function.

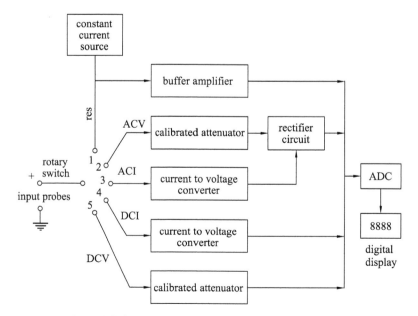

Figure 2-2-4 The block diagram of a digital multimeter

(1) The 1st position of the switch is to measure the resistance.

(2) The 2nd position of the switch is to measure the AC voltage.

(3) The 3rd position of the switch is to measure the AC current.

(4) The 4th position of the switch is to measure the DC current.

(5) The 5th position of the switch is to measure the DC voltage.

The front panel of DMM consists of display, selection knob and ports. The main functions are briefly given as follows.

(1) *Display*: The display parts typically occupies a position at the top of the instrument. The display on the common DMM have four digits, where the first digit (called half digital) is only 0 or 1 and the other 3 digits (called full digital) is from 0-9. In addition, symbol " +/ - " can also be displayed on the DMM.

(2) *Main ports*: "Common" is connected to the negative or black lead and probe; "Volts and ohms" is connected to the positive or red lead and probe; "Amps" is connected to the red lead and probe.

(3) *Selection knob*: The selection knob typically occupies a position at the center of the instrument. It is used to select the type of measurement to be performed and the required range.

The operation of the DMM is usually very straightforward. If the meter is new, it is obvious that a battery must be installed to power it. See the DMM's operating instructions for details. With a knowledge of how to measure voltage, current and resistance, you are ready to use the DMM.

Using the DMM, the steps of measuring the voltage are briefly given as follows:

(1) Switch on the digital multimeter.

(2) Insert the black test probe into the COM input jack and the red test probe into the "V" input jack.

(3) If the DMM has only manual ranging, select the maximum range to avoid input overload.

(4) Connect the probe tips to the circuit across a load or power source (in parallel to the circuit).

(5) Record the readings and don't forget the units of measurement.

Using the DMM, the steps of measuring the DC and AC current are briefly given as follows:

(1) Switch off the circuit power.

(2) Select A ~ (AC) or A (DC) as desired.

(3) Insert the black test probe into the COM input jack and the red test probe into the amp input jack.

(4) Connect the probe tips to the circuit across the slots so that the current will flow through the DMM (series connection).

(5) Switch on the circuit power.

(6) Record the readings and don't forget the units of measurement.

Using the DMM, the steps of measuring the resistance are given as follows:

(1) Switch on the digital multimeter.

(2) Insert the black test probe into the COM input jack and the red test probe into the Ω input jack.

(3) If the DMM has only manual ranging, select the maximum range to avoid input overload.

(4) Connect the probe tips across the component or portion of the circuit.

(5) Record the readings and don't forget the units of measurement, such as ohms (Ω), kilo-ohms (kΩ), mega-ohms (MΩ).

After the reading is completed, a wise precaution is to place the probe into a voltage measurement socket and convert the range to the maximum voltage. In this way, if the DMM is accidentally connected without considering the range of use, the chance of damaging the meter is small.

2.2.4 Electrodynamo watt meters

The block diagram of an electrodynamo watt meter is illustrated in Figure 2-2-5. An electrodynamometer type watt meter depends on the reaction between the magnetic field of moving and fixed coils. The internal structure of a watt meter consists of two coils. The fixed coil connected in series with the circuit is called the current coil which is made of thick gauge copper wire and has lower resistance. To simplify the structure, the fixed coil is divided into two parts. On the other hand, the moving coil connected in parallel with the circuit is called the pressure coil which is made of a thin gauge wire and has relatively high resistance. Electrodynamo watt meters are used to measure the power of the AC and DC circuits.

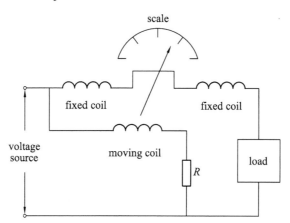

Figure 2-2-5 The block diagram of an electrodynamo watt meter

Let the root mean square voltage of circuit is V and current flowing is I. As a high value of series resistance is connected in series with the pressure coil, pressure coil circuit can be assumed purely resistive. Therefore, the current flowing through the pressure coil is in phase with the circuit voltage V. The current of the pressure coil is given as

$$i_p = \frac{\sqrt{2}V}{R}\sin \omega t \qquad (2\text{-}2\text{-}2)$$

If the circuit current is assumed lagging by an angle φ, then the current through the current coil is given as

$$i_c = \sqrt{2} I \sin(\omega t - \varphi) \quad (2\text{-}2\text{-}3)$$

Instantaneous deflecting torque in electrodynamo meter type instrument is given as

$$T_d = i_p i_c \frac{dM}{d\theta} = [\cos \varphi - \cos(2\omega t - \varphi)] \frac{VI}{R} \frac{dM}{d\theta} \quad (2\text{-}2\text{-}4)$$

where M is the mutual inductance between fixed and moving coils, θ is the deflection angle of the controlling torque due to spring. As the instrument will respond to the average deflecting torque over a time period, the average of deflecting torque is given as

$$\overline{T}_d = \frac{VI \cos \varphi}{R} \frac{dM}{d\theta} \quad (2\text{-}2\text{-}5)$$

As the spring provide the controlling torque, the controlling torque T_c is given as

$$T_c = K\theta \quad (2\text{-}2\text{-}6)$$

where K is spring constant and θ is final steady deflection angle.

When the controlling torque is equal to the average of deflecting torque, the pointer of electrodynamo meter is at the balanced position,

$$T_c = \overline{T}_d; \theta = \frac{VI \cos \varphi}{KR} \frac{dM}{d\theta} = \frac{P}{KR} \frac{dM}{d\theta} \quad (2\text{-}2\text{-}7)$$

The Equation 2-2-7 shows the angular deflection of electrodynamo meter is proportional to the active power. The shape of scale of this instrument depends on the variation of $dM/d\theta$, which means the variation of mutual inductance between the fixed coil and moving coil. By properly designing the fixed and moving coil, $dM/d\theta$ can be made to vary linearly over a range of 40° – 50° on either side of the zero mutual inductance.

The errors in the electrodynamo watt meter are given as follows.

(1) Inductance and capacitance of moving coil.
(2) Mutual inductance effects between the moving coil and fixed coil.
(3) Eddy currents and stray magnetic field.
(4) Vibration of moving system and temperature error.

2.2.5 Spectrum analyzers

Spectrum analyzers are widely used instruments for radio frequency signal spectrum, electromagnetic interference and electromagnetic compatibility. Spectrum analyzers are classified into filter bank spectrum analyzer, super-heterodyne spectrum analyzer, digital FFT spectrum analyzer, real time spectrum analyzer, PXI spectrum analyzer, and USB spectrum analyzer.

A super-heterodyne spectrum analyzer is considered as a form of radio receiver with a display at the output level, which the receiver is tuned or scanned over the required range and the filters are chosen to accept the required signal bandwidth. The block diagram is illustrated in Figure 2-2-6. The radio frequency signal is applied to be attenuated by an input attenuator, and then to go through low pass filter. The output of low pass filter and voltage tuned oscillator signal are applied to the mixer. The output signal of mixer is the difference frequencies of the two-input signal, also called intermediate frequency (IF) signal. After IF amplifier, the detector can only acquire the amplitude of IF signal and discard the phase information. The voltage tuned oscillator is controlled by ramp generator. As a result, ramp generator can be displayed by frequency. Finally, the RF signal spectrum is displayed on cathode ray oscilloscope (CRO) screen.

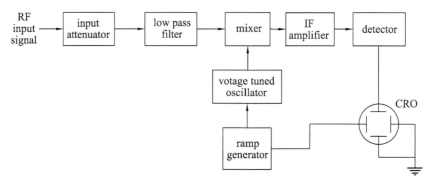

Figure 2-2-6 The block diagram of a super-heterodyne spectrum analyzer

The super-heterodyne spectrum analyzers are able to operate over wide frequency range and have wide bandwidth. However, the disadvantages of super-heterodyne spectrum analyzer are its inability to measure RF phase and capture transient events.

The block diagram of a filter bank spectrum analyzer is illustrated in Figure 2-2-7. It has a set of band pass filters (filter 1, filter 2, ⋯, filter n) and each one is designed for allowing a specific band of frequencies. The output of each band pass filter is provided to a corresponding detector (detector 1, detector 2, ⋯, detector n). All detectors' outputs are connected to electronic switch. This switch allows the detector outputs sequentially to the vertical deflection plate of CRO. As a result, CRO displays the frequency spectrum of AF signal on its CRO screen. Filter bank spectrum analyzer is similar to super-heterodyne spectrum analyzer, it cannot measure the phase of RF signal.

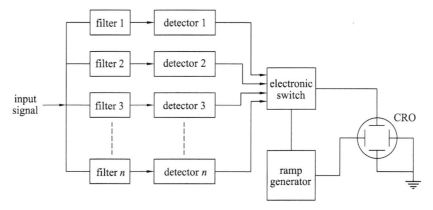

Figure 2-2-7 The block diagram of a filter bank spectrum analyzer

Digital FFT spectrum analyzer uses Fourier analysis and digital signal processing techniques to provide spectrum analysis. The Fourier transform is developed by Joseph Fourier (1768—

1830), which is suitable for the continuous time domain. The Fourier transform of continuous time function $f(t)$ is given as

$$F(j\omega) = \int_{-\infty}^{\infty} f(t) e^{-j\omega t} dt \qquad (2\text{-}2\text{-}8)$$

The Equation 2-2-8 is the spectrum of continuous time function $f(t)$, which includes both magnitude and phase information. Similarly, the discrete Fourier transform (DFT) is suitable for the discrete time domain. The DFT of discrete time sequence $x(n)$ is given as

$$X(k) = \sum_{n=0}^{N-1} x(n) e^{-j2\pi \frac{kn}{N}} \qquad (2\text{-}2\text{-}9)$$

The Nyquist theorem states that as long as the sampling rate is greater than twice the highest frequency component of the signal, the sampled data will accurately represent the input signal. Therefore, the spectrum of continuous time signal can be calculated by DFT. To achieve greater resolution, a large number of samples in the discrete time domain is required. As a result, the direct computation of the DFT is inefficient. As well known, the Fast Fourier transform (FFT) is an efficient algorithm for the computation of the DFT. The FFT requires that the number of samples in the time domain should be equal to an integer power of 2.

The block diagram of a digital FFT spectrum analyzer is illustrated in Figure 2-2-8. If the input analogue signal level is too high or too low, then distortion or noise will be occurred. The analogue front-end attenuator/gain can provide either gain/attenuation in order to ensure that the signal is at the required level for the analogue to digital conversion (ADC). To avoid signal aliasing, a low pass filter is placed in front of the sampler to remove any unwanted high-frequency signal. Through sampling and ADC stages, the analogue signal is converted into the digital signal. Using the digital FFT techniques, the data in the time domain are transformed into the frequency domain. Finally, the display elements can present the spectrum information with all kinds of flexible formats, such as logarithmic and linear.

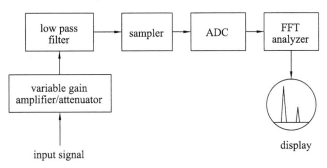

Figure 2-2-8　The block diagram of a digital FFT spectrum analyzer

Before using a spectrum analyzer, you need to understand its specifications. Some basic specifications of spectrum analyzers are briefly given as follows.

(1) *Frequency reference inaccuracy*: This specification defines that the frequency reference inaccuracy is chiefly determined by the internal oscillator in the analyzer.

(2) *Marker resolution*: The marker resolution is not actually related to the frequency

accuracy, but the marker can make it the step size between one position and the adjacent one.

(3) *Absolute amplitude accuracy*: This specification refers to measurements where the absolute level is required.

(4) *Relative amplitude accuracy*: This specification is used when signals are expressed in terms of decibels compared to another signal.

(5) *Resolution bandwidth*: The resolution bandwidth is primarily determined by the bandwidth of the filter within the analyzer, but other factors like filter type and noise sidebands are factors to be taken into consideration.

(6) *Phase noise*: It is typically given as the single sideband noise level, which is measured with a perfect signal source. The typical phase noise of spectrum analyzer is listed in Table 2-2-1.

Table 2-2-1 The typical phase noise

offset from carrier	level
10 Hz	−80 dBc
100 Hz	−108 dBc
1 kHz	−125 dBc
10 kHz	−135 dBc
100 kHz	−138 dBc
1 MHz	−145 dBc
10 MHz	−154 dBc

2.2.6 AC voltmeters

Measuring RF voltage is not an easy task because its amplitude is very small and its frequency is high. To measure small RF voltage signals, small voltage signals should be amplified. In addition, a permanent magnet moving coil (PMMC) meter is only driven by the direct current. Input AC signal needs to be converted to DC by a rectifier. As a result, the AC millivolt meter consists of an amplifier and a rectifier. As shown in Figure 2-2-9, there are rectifier amplifier type and amplifier rectifier type AC voltmeter. The voltage level of input AC signal is adjusted by attenuator. The output of attenuator is provided to rectifier which can convert the AC voltage into pulsating DC voltage. The rectifier can be used before or after the multi-stage amplifier. It depends on the AC/DC type of amplifier. The rectifier circuit will be used after the multi-stage AC amplifier. Conversely, the rectifier circuit will be used before the multi-stage DC amplifier. Finally, PMMC meter displays the reading.

According to the rectifier type, there are average reading type, peak reading type and true RMS reading type AC voltmeter.

(1) *Average reading AC voltmeter*: the meter of the average reading AC voltmeter is calibrated in terms of the RMS values for a sine wave. The form factor is defined as the rms value divided by the average value,

$$\alpha = \frac{U_\sim}{\overline{U}} \tag{2-2-10}$$

where $U_\sim$ is the RMS of an AC signal, $\overline{U}$ is the average value of an AC signal. For a pure sinusoidal waveform, the form factor is equal to 1.11. The form factor for rectangle waveform is equal to 1. The form factor for a triangle waveform is equal to 1.15.

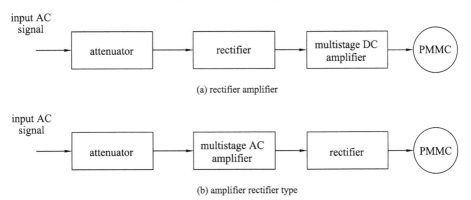

Figure 2-2-9 The block diagram of an AC voltmeter

In the case of non-sinusoidal waveform voltage, the RMS value of non-sinusoidal waveform voltage is expressed by

$$U_{non} = 0.9\alpha U_{read} \tag{2-2-11}$$

where U_{non} is the RMS value of non-sinusoidal waveform voltage, α is the form factor, and U_{read} is the reading value.

(2) *Peak reading AC voltmeter*: the meter of the peak reading AC voltmeter is calibrated in terms of the RMS values for a sine wave. The crest factor is defined as the peak value divided by RMS value,

$$\beta = \frac{U_P}{U_\sim} \tag{2-2-12}$$

where U_P is the peak value of an AC signal. The crest factor for a sinusoidal waveform is 1.414. The crest factor for a rectangle waveform is 1. The crest factor for a triangle waveform is 1.72. In the case of non-sinusoidal waveform voltage, the RMS value of non-sinusoidal waveform voltage is expressed by

$$U_{non} = \frac{\sqrt{2}}{\beta} U_{read} \tag{2-2-13}$$

(3) *True RMS reading of AC voltmeter*: In order to measure the rms voltage of non-sinusoidal waveforms, a true RMS reading voltmeter is required. The principle of a true rms reading voltmeter is to detect the heating power of a non-sinusoidal waveform. It is well known that heating power is proportional to the square of the root mean square voltage. As shown in Figure 2-2-10, a true rms reading voltmeter consists of main thermocouple, balancing thermocouple, AC amplifier and DC amplifier. After the non-sinusoidal waveform is amplified by the AC amplifier,

the main thermocouple measures the heating power of the non-sinusoidal waveform. In addition, a balanced thermocouple was introduced to eliminate the nonlinear effects of the main thermocouple.

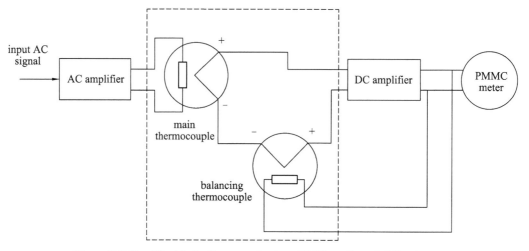

Figure 2-2-10 The block diagram of a true RMS reading of AC voltmeter

2.3 Measurement basis

2.3.1 Significant figures

Significant figures indicate the accuracy of the measurement, also called significant digits. As well known, a number with more significant digits is more precise. As a result, you should understand the rules about significant digits when recording a measurement data. The rules about significant digits are briefly given as follows.

(1) All non-zero numbers are significant.

(2) Zeros between non-zero numbers are significant.

(3) Leading zeros before a number are not significant.

(4) Unless there is a decimal point, the zeros after the number are not significant.

There are currently different rounding conventions in use. In measured values, 5 is rounded to the nearest even number. It is necessary to round the results to the correct number of significant figures in calculating measurement data. The rules are briefly given as follows.

(1) When make addition or subtraction, the number of decimal places in the result is equal to the number of decimal places in the least precise value.

(2) When make multiplication or division, the number of significant digits in the result is equal to the smallest number of significant digits of the value.

(3) When involve multiple arithmetic operations, the number of significant figures is

determined by the rules in the following order: i. operation in parentheses; ii. multiplication; iii. division; iv. addition; v. subtraction.

Significant figures are used to provide numerical accuracy. Although the rounding process involved still introduces a measure of error into the numbers, the correct use of significant figures will be sufficient to maintain the required level of precision.

2.3.2 Measurement

Measurement is generally referred as a laboratory process to determine a physical variable with instruments and express the results in numbers, curve, and tabular formats with physical unit. As shown in Figure 2-3-1, the measurement system consists of the measurement object, instruments, measurement principle and technique, staff member and measurement environment.

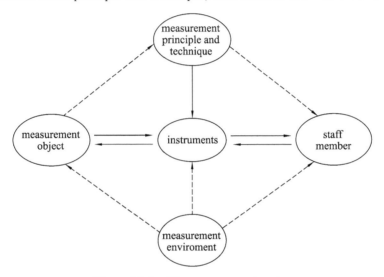

Figure 2-3-1 The measurement system

Measurement environment refers to any data collection activity involving the assessment of chemical, physical, and biological factors in the environment which affect human health. The objective of the measurement is to obtain the true value of an unknown physical variable. The "true value" typically referred to be obtained in an ideal measurement with no error. In other words, the "true value" is perfectly accurate.

However, no measurement can be made with perfect accuracy. Measurement error is the difference between a measured quantity and its true value. As absolute error is the difference between the measured value of a quantity and its true value, it has unit. The absolute error is defined as

$$\Delta x = x - x_0 \tag{2-3-1}$$

where x means the measured value, x_0 means the true value.

Relative error represents the ratio of the absolute error of the measurement to the true value. As the relative error is expressed as a percentage or as a fraction, it has no units. The relative error is defined as

$$\gamma = \frac{\Delta x}{x_0} \times 100\% \qquad (2\text{-}3\text{-}2)$$

Calibration error indicates the quality of the measuring instruments made by the manufacturer. Calibration error is usually expressed as a percentage of reading or as a percentage of full-scale deflection. The span is defined as the range of an instrument from the minimum to maximum scale value. If the minimum of an instrument is zero, then the span is its full scale. As shown in Figure 2-3-2, the span error is the algebraic difference between the indication and the actual value of the measured variable. The formula of percentage span error is defined as

$$\gamma_s = \frac{x_r - x_0}{x_s} \times 100\% \qquad (2\text{-}3\text{-}3)$$

where x_s means the span of the instrument, x_r means the readings or output of the instrument, γ_s means the percentage span error.

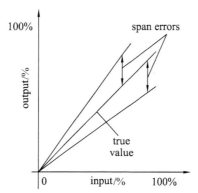

Figure 2-3-2 The span errors

2.3.3 Measurement error types

According to the properties of measurement errors, there are chiefly systematic error, random error and gross error. A systematic error is not determined by chance but is introduced by inaccuracies in the system inherent, such as offset, zero setting error, multiplier, and scale factor error. In other words, a systematic error is consistent, repeatable error associated with faulty equipment or a flawed experiment design. As a result, the most direct way to eliminate systematic errors is to use calibrated equipment.

Random errors in experimental measurements are caused by unknown and unpredictable changes in the experiment. These changes may occur in the measuring instruments or in the environmental conditions, for example, electronic noise in the circuit of an electrical instrument.

In such cases, statistical methods may be used to analyze the data. The central limit theorem states that random variable n matter what the shape of the distribution, the sampling distribution of the sample means approaches a normal distribution as the sample size gets larger. According to the central limit theorem and the causes of random errors, random errors often have a Gaussian or normal distribution.

Chapter 2 Electronic Components, Instruments, and Measurement Basis

The formula of the sample mean is given as

$$\overline{x} = \frac{1}{N}\sum_{i=1}^{N} x_i \qquad (2\text{-}3\text{-}4)$$

where $\overline{x}$ means sample mean, x_i means i-th items in the sample, N means the number items in the sample. The variance refers to how spread out samples is around the mean. In other words, the variance shows the accuracy of the estimate. The variance of the sampling distribution of the mean is expressed by

$$\sigma^2(\overline{x}) = \sigma^2\left(\frac{1}{N}\sum_{i=1}^{N} x_i\right) = \frac{1}{N}\sigma^2 \qquad (2\text{-}3\text{-}5)$$

where σ^2 means the variance of the sample. The Equation 2-3-5 shows that the variance of the sample mean is equal to one Nth of sample variance. As N grows larger, the variance or standard deviation becomes smaller. In other words, random errors can be reduced by repeating measurement multiple times. The relationship between systematic error and random error is illustrated in Figure 2-3-3.

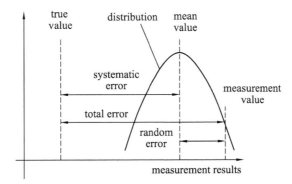

Figure 2-3-3 The relationship between systematic error and random error

To characterize the measurement uncertainty, the following terminology are often used:

(1) *Accuracy*: Accuracy refers to the error between the real value and the measured value, which is used to characterize the systematic error.

(2) *Precision*: Precision refers to the random spread of measured values around the average measured values, which is used to characterize the random error.

(3) *Resolution*: The smallest error is distinguished magnitude from the measured value. The specified resolution of an instrument has no relation to the accuracy of measurement.

Compared to systematic errors, random errors are unpredictable and unavoidable, and they can't be replicated by repeating the experiment again. As shown in Figure 2-3-4, the main difference between systematic and random errors is that random errors lead to fluctuations around the true value, whereas systematic errors lead to predictable and consistent departures from the true value due to problems with the instrument calibration.

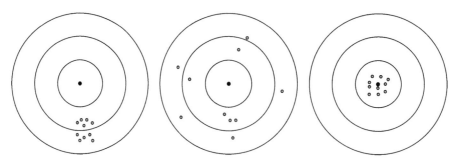

Figure 2-3-4 The measurement uncertainty

The gross errors mainly cover human mistakes, such as misreading of instruments, incorrect adjustment of instruments, and computational mistakes. Whenever human beings are involved, some gross errors will inevitably occur. However, some gross errors are easily detected and deleted with threshold methods.

2.3.4 AD converters

An analog to digital converter (ADC) converts an analog signal into a digital signal. The digital signal is represented by a binary code. The ADC is a linkage between the analog world of transducers and discreet world of the data. It can be divided into a counter type ADC, a successive approximation ADC, a flash type ADC, and a dual slope integration type ADC. The ADCs are widely used in transducer, cell phones, micro-controllers, DMM, and digital storage oscilloscopes.

The block diagram of the successive approximation ADC is illustrated in Figure 2-3-5. The analog signal level is adjusted by input attenuator. In order to maintain the ADC accuracy, the sample and hold circuit is introduced to fix the value of input analog voltage within a specified minimum time period. When the control logic receives the start command signal, it resets all bits of the successive approximation register (SAR) and enables the clock signal generator to send clock pulses to the SAR.

From the most significant bit to least significant bit, DAC converts the received digital input from SAR into an analog output. The comparator compares this analog value from DAC with the external analog input signal and determines the state of the current bit according to the binary search method.

The SAR logic outputs a binary code to the DAC that is dependent on the current bit under scrutiny and the previous bits already approximated. Once all bits have been approximated, the digital approximation is output at the end of the conversion. Thus, the conversion time of successive approximation ADC is proportional to the number of bits of the digital output.

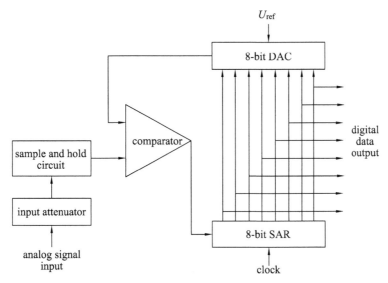

Figure 2-3-5 The block diagram of a successive approximation ADC

In a dual-slope type ADC as shown in Figure 2-3-6, an integrator is used to integrate with an unknown analog input voltage U_{in} for a fixed period of time T_1. Thus, the output voltage of the integrator is expressed by

$$U_{o1} = -\frac{1}{RC}\int_0^{T_1} U_{in}(t)\,dt = -\frac{1}{RC}\overline{U}_{in}T_1 \qquad (2\text{-}3\text{-}6)$$

$$\overline{U}_{in} = \frac{1}{T_1}\int_0^{T_1} U_{in}(t)\,dt \qquad (2\text{-}3\text{-}7)$$

where $\overline{U}_{in}$ is the average of an unknown analog input voltage, R is the resistance of the integrator, C is the capacitive of the integrator.

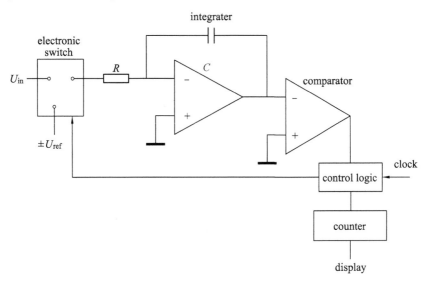

Figure 2-3-6 The block diagram of a dual slope ADC

To integrate with a known reference voltage U_{ref} by the same integrator, the output voltage after the second integral is zero, which can be expressed by

$$U_{o2} = U_{o1} + \left(-\frac{1}{RC}\int_0^{T_2} U_{ref}dt\right) = 0 \quad (2\text{-}3\text{-}8)$$

Combined the Equation 2-3-7 with the Equation 2-3-8, the unknown analog input voltage is expressed by

$$\overline{U}_{in} = \frac{T_2}{T_1}U_{ref} \quad (2\text{-}3\text{-}9)$$

The Equation 2-3-9 shows that the dual-slope ADC will produce a value which is equal to the average of the unknown input analog signal. If the unknown analog input voltage U_{in} is constant, then $\overline{U}_{in}$ is equal to U_{in}. As shown in Figure 2-3-7, the output waveform of the dual-slope ADC is the dual ramp. The characteristics of the dual slope ADC is briefly given as follows:

(1) The voltage measurement is converted to time measurement, so it requires the counter.

(2) Although it requires the integrator, the conversion accuracy is independent of the resistor and capacitor of the integrator.

(3) The conversion speed is not high, because it requires two integrals.

(4) There is no need to provide a sample and hold circuit, because changes in the input voltage will not cause significant errors.

(5) It is highly reliable and effective when used with noisy signals, because the noise can be reduced by the average.

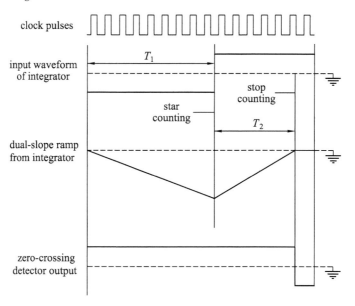

Figure 2-3-7 The waveform of a dual slope ADC

In order to proper use the instruments and analyze the errors, you need to understand some basic specifications of an AD converter which are briefly given as follows.

(1) *Resolution*: For a N bits AD converter, it is capable of producing distinct output codes 2^{-N}.

(2) *Accuracy*: The accuracy is determined by its internal circuit and noise from external sources connected to the A/D input.

(3) *Offset error*: It is defined as a deviation of the code transition points that is present across all output codes.

(4) *Gain error*: It determines the amount of deviation from the ideal slope of the A/D converter transfer function.

(5) *Differential non-linearity* (*DNL*): This specification defines the deviation of the real code width from the ideal code width.

(6) *Integral non-linearity* (*INL*): This specification defines the deviation of the measured transfer curve from the adjusted transfer curve.

(7) *Absolute error*: It is also called the total unadjusted error. It is the sum of the offset, gain, differential non-linearity errors and integral non-linearity errors.

2.3.5 DA converters

A digital to analog converter (DAC) is a device that transforms digital data into analog signal. According to the Nyquist-Shannon sampling theorem, a DAC can reconstruct sampled data into analog signal with precision. The digital data are used in computers, microprocessors, integrated circuits, and mobile devices. In these applications, the digital data requires the conversion to an analog signal in order to link with the real world. The DAC is divided into binary weighted resistor DAC, R-2R ladder DAC, serial DAC, bipolar DAC.

As shown in Figure 2-3-8, a 4-bit binary weighted resistor DAC consists of a set of switches, parallel binary weighted resistor band, reference voltage source and op-amp. Each bit of the input binary data is determined by the switch and the op amp is used to the current sum of the resistor bank which is proportional to the digital input. If the feedback resistor R_f is equal to R, then the op amp output is expressed by

$$U_o = - U_{ref}[b_0 2^{-1} + b_1 2^{-2} + \cdots + b_{n-1} 2^{-n}] \tag{2-3-10}$$

where n is the DAC bits and U_{ref} is the reference voltage source and the $b_0, b_1, \cdots, b_{n-1}$ means the input digital data.

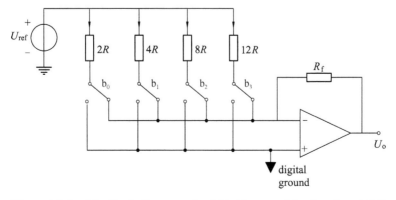

Figure 2-3-8 The block diagram of a 4-bit binary weighted resistor DAC

When the number of binary input increases, it is difficult to maintain the relationship of the resistor's ratio. For high accuracy of DAC applications, the values of resistances must be accurate, which depends primarily on the temperature.

As shown in Figure 2-3-9, only two values of resistors are required in the 4-bit R-$2R$ ladder DAC. Therefore, it does not require high precision resistors. In addition, the reference voltage can affect the accuracy of DAC. Compared to binary weighted resistance DAC, it has lower conversion speed which is determined by clock speed of the input signal and settling time of the DAC. The settling time requires for the output signal to settle within $\pm 1/2$ LSB of its final value after a given change in input scale. As shown in Figure 2-3-10, the settling time include the slew time, recovery time and linear recovery time, which is limited by slew rate of output amplifier.

Then the op amp output is expressed by

$$U_o = -U_{ref}\frac{R_f}{2^n R}[b_0 2^{n-1} + b_1 2^{n-2} + \cdots + b_{n-1} 2^0] \qquad (2\text{-}3\text{-}11)$$

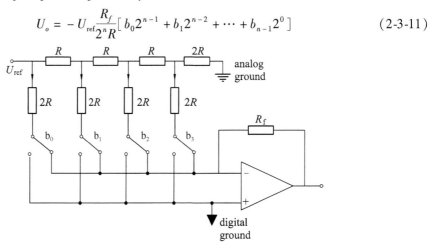

Figure 2-3-9 The block diagram of a 4-bit R-$2R$ ladder DAC

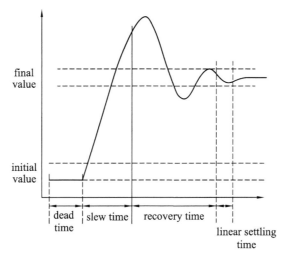

Figure 2-3-10 The settling time of a DAC

In DAC devices, some typical errors consist of gain error, offset error, full scale error, resolution error, non-linearity error, non-monotonic error, settling time and overshoot error. The

Chapter 2 Electronic Components, Instruments, and Measurement Basis

gain error is illustrated in Figure 2-3-11a and the offset error is illustrated in Figure 2-3-11b. The full scale error consists of gain error and offset error.

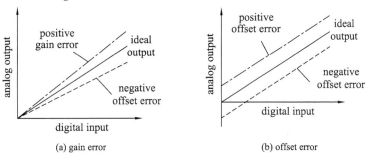

Figure 2-3-11 a DAC errors

As shown in Figure 2-3-12, non-linearity errors include DNL error and INL error. The DNL error is a way to characterize the difference between two successive voltage level produced by a DAC. The DNL is the maximum deviation of the output steps from the ideal analog LSB value, which has the positive or negative properties. The INL error is defined as the maximum deviation between the actual input-output characteristics and the ideal transfer characteristics.

By calculating the algebraic sum of all previously converted DNL error, the INL error is expressed by

$$\varepsilon_{INL} = \sum_{i=0}^{n} \mid \varepsilon_{DNL,i} \mid \qquad (2\text{-}3\text{-}12)$$

where ε_{INL} means the INL error, n means the bits of a DAC, and $\varepsilon_{DNL,i}$ means the DNL error of the DAC i-th bit.

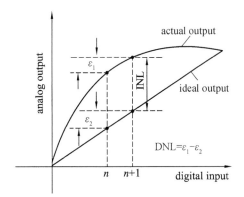

Figure 2-3-12 The DNL and INL error of a DAC

Chapter 3

Electrical Engineering Lab Parts

3.1 Voltage-current characteristic

【Lab objective】

(1) Understand the voltage-current characteristic of resistors and silicon diodes.
(2) Learn to measure voltage and current of resistors and silicon diodes.
(3) Learn to use an oscilloscope and a function generator.

【Lab devices】

Resistors, regulated DC power supply, DC ammeter and voltmeter, function generator, oscilloscope, silicon diode, and wires.

【Lab principle】

A linear resistor follows Ohm's law and has a constant resistance. The Ohm's law is defined as
$$U = RI \qquad (3\text{-}1\text{-}1)$$
where R is resistance, I is the current flowing through the resistor, U is the voltage of the resistor. As shown in Figure 3-1-1, the voltage-current characteristic of a linear resistor is a straight line passing through the origin. In addition, a nonlinear resistor does not follow Ohm's law.

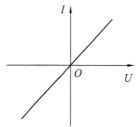

Figure 3-1-1 The voltage-current characteristic of a linear resistor

A diode is defined as a two-terminal electronic component that only conducts current in one direction which is operated within a specified voltage level. A diode will have negligible resistance in one direction to allow current to flow through it which is called "forward biased", and a very high resistance in the reverse direction to prevent current from flow through it which is called "reverse biased". A diode is effectively like a valve for an electrical circuit.

A PN junction is the simplest form of the semiconductor diode. In ideal conditions, this PN junction behaves as a short circuit when it is forward biased, and as an open circuit when it is the reverse biased. The volt-ampere characteristic of the PN-junction diode is a curve between the voltage of the junction and the circuit current. The characteristic curve of PN-junction diode is designed under forward bias and reverse bias. The voltage-current characteristic curve of a diode is entirely non-linear. Typically, a silicon diode will have a forward bias voltage around 0.6 ~ 1 V. A germanium-based diode might be lower, around 0.3 V, which is illustrated in Figure 2-1-6.

【Lab procedure】

(1) *Voltage-current characteristic of a linear resistor*: Connect the circuit as shown in Figure 3-1-2, and switch on the DC power supply after checking the circuit. The resistance value of the resistor is 1 kΩ. Measure a set of voltages and currents of the resistor by rotating the knob of voltage source. Make records of the measured values and fill them in the Table 3-1-1. Draw the voltage-current characteristic curve on the graph sheet.

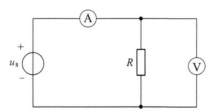

Figure 3-1-2 Resistor circuit

Table 3-1-1 Voltage and current of resistor

u_s/V	0	3	6	9	12	15
U_R/V						
I/mA						

(2) *Voltage-current characteristic of a silicon diode*: Connect the circuit as shown in Figure 3-1-3, and switch on the DC power supply after checking the circuit. The resistance value of the resistor is 1 kΩ and diode model is 1N4007. Measure a set of voltages and currents of the diode by rotating the knob of voltage source. Make records of the measured values and fill them in the Table 3-1-2. Draw the voltage-current characteristic curve on the graph sheet.

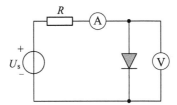

Figure 3-1-3   Voltage-current characteristic of a silicon diode

Table 3-1-2   Voltage and current of diode

U_{diode}/V	0	0.2	0.4	0.6	0.62	0.65	0.68	0.7
I/mA								

Connect the circuit as shown in Figure 3-1-4, and switch on the function generator after checking the circuit. AC voltage signal is a sinusoidal wave with a frequency of 1 kHz and the maximum amplitude of 0.5 ~ 2.5 V, which is from the function generator. Observe the output waveforms across the diode using an oscilloscope and draw the waveforms across the diode.

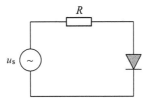

Figure 3-1-4   Silicon diode circuit under the AC voltage

【Pre-lab questions】

(1) Preview the voltage-current characteristic of linear resistors.
(2) Preview the voltage-current characteristic of diodes.
(3) Preview the forward biased and the reverse biased features of diodes.

【Post-lab questions】

(1) What is the difference between linear and non-linear resistors?
(2) What is the voltage-current characteristic of a diode?
(3) How to measure the breakdown voltage of a silicon diode?
(4) How to identify the positive pole or negative pole of a diode?

Chapter 3 Electrical Engineering Lab Parts

3.2 Kirchhoff's law and superposition principle

[Lab objective]

(1) Understand the reference direction of voltage and current in the circuits.
(2) Understand the Kirchhoff's law and superposition principle.
(3) Learn to verify Kirchhoff's law and superposition principle using the lab method.

[Lab devices]

Resistors, regulated DC power supply, DC ammeter and voltmeter, and wires.

[Lab principle]

Kirchhoff's laws were first introduced in 1847 by the German physicist Gustav Robert Kirchhoff. Kirchhoff's law is based on the law of conservation of charge, which requires that the algebraic sum of charges within a system cannot change. Kirchhoff's current law is expressed by

$$\sum_{n=1}^{N} i_n = 0 \qquad (3\text{-}2\text{-}1)$$

Kirchhoff's current law states that the sum of the currents entering a node is equal to the sum of the currents leaving the node. As shown in Figure 3-2-1a, the current of the node is equal to

$$I_1 + I_2 + I_4 - I_3 - I_5 \qquad (3\text{-}2\text{-}2)$$

Kirchhoff's voltage law states that the algebraic sum of all voltages around a closed path is zero. Kirchhoff's voltage law is expressed by

$$\sum_{n=1}^{N} v_n = 0 \qquad (3\text{-}2\text{-}3)$$

As shown in Figure 3-2-1b, the voltage of loop is equal to

$$U_{ab} + U_{bc} + U_{cd} + U_{da} = 0 \qquad (3\text{-}2\text{-}4)$$

The superposition principle states that the voltage across an element in a linear circuit is the algebraic sum of the voltages across that element due to each independent source acting alone.

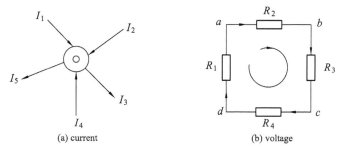

(a) current (b) voltage

Figure 3-2-1 Kirchhoff's laws

[Lab procedure]

(1) *Kirchhoff's laws*: Notice the reference direction of current I_1, I_2 and I_3. There are two voltage sources U_{s1} and U_{s2} and two switches S_1 and S_2. Let switch S_1 connect to the voltage source U_{s1} and switch S_2 connect to the voltage source U_{s2}. The voltage source U_{s1} and U_{s2} work simultaneously. Connect the circuit as shown in Figure 3-2-2, and switch on the DC power supply after checking the circuit. Observe the results, make records of the measured values, and fill them in Table 3-2-1.

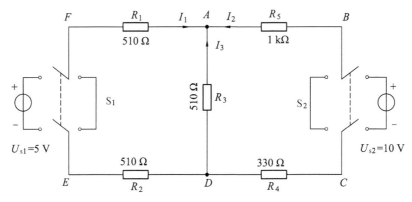

Figure 3-2-2 The circuit for Kirchhoff's law and the superposition theorem

Table 3-2-1 Kirchhoff's laws

items	I_1/A	I_2/A	I_3/A	U_{EF}/V	U_{CB}/V	U_{FA}/V	U_{AB}/V	U_{AD}/V	U_{CD}/V	U_{DE}/V
measured value										
theoretical value										

(2) *Superposition principle*: Let switch S_1 connect to the voltage source U_{s1} and switch S_2 connect to wire, the voltage source U_{s1} works only. Connect the circuit as shown in Figure 3-2-3a, and switch on the DC power supply after checking the circuit. Observe the results, make records of the measured values, and fill them in Table 3-2-2.

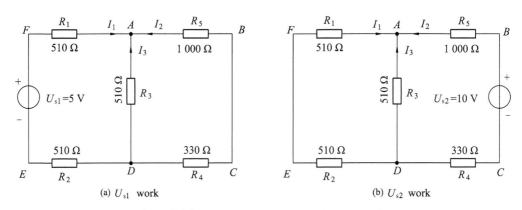

(a) U_{s1} work (b) U_{s2} work

Figure 3-2-3 The voltage source works separately

Table 3-2-2 Experimental data and theoretical value

items	I_1/A	I_2/A	I_3/A	U_{AD}/V
measured value (U_{s1} work only)				
theoretical value (U_{s1} work only)				
measured value/(U_{s2} work only)				
theoretical value/(U_{s2} work only)				

Similarly, let switch S_1 connect to wire and switch S_2 connect to the voltage source U_{s2}, the voltage source U_{s2} works only. Connect the circuit as shown in Figure 3-2-3b, and switch on the DC power supply after checking the circuit. Observe the results, make records of the measured values, and fill them in Table 3-2-2.

【Pre-lab questions】

(1) Preview Kirchhoff's law and Ohm's law.
(2) Preview the superposition principle.
(3) Preview the reference direction of voltage and current.
(4) Preview the concept of nodes and loops in the circuits.

【Post-lab questions】

(1) How to verify the Kirchhoff's law using the voltage and current data?
(2) How to verify the superposition principle using the voltage and current data?
(3) Can superposition principle be applied to the diodes?
(4) How to calculate the power of resistor R_3? Is it satisfied the superposition principle and Kirchhoff's law?

3.3 Thevenin's theorem

【Lab objective】

(1) Understand the principle of Thevenin's theorem.
(2) Understand the open-circuit voltage and the equivalent resistance.
(3) Learn to verify Thevenin's theorem using the lab method.

【Lab devices】

Resistors, regulated DC power supply, DC ammeter and voltmeter, and wires.

【Lab principle】

Thevenin's theorem states that a linear two-terminal circuit can be replaced by an equivalent circuit consisting of a voltage source U_{Th} in series with a resistor R_{Th}, where U_{Th} is the open-circuit voltage at the terminals and R_{Th} is the input or equivalent resistance at the terminals. The diagram of replacing a linear two-terminal circuit by its Thevenin equivalent circuit is illustrated in Figure 3-3-1.

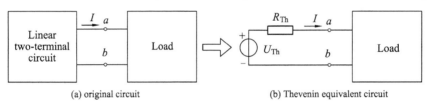

(a) original circuit   (b) Thevenin equivalent circuit

Figure 3-3-1 Replacing a linear two-terminal circuit by its Thevenin equivalent

There are many measurement methods to achieve the Thevenin equivalent resistance of linear two-terminal circuit. Two commonly used methods are briefly given as follows.

(1) *Open-circuit voltage and short-circuit current*: The open-circuit voltage can be directly measured by using the DC voltmeter. Short-circuit current is measured by using ammeter. Therefore, the equivalent resistance at the terminals will be calculated by

$$R_0 = \frac{U_{oc}}{I_{sc}} \qquad (3\text{-}3\text{-}1)$$

where U_{oc} is open-circuit voltage and I_{sc} is short-circuit current.

(2) *Voltage-current method*: Voltage-current method is illustrated in Figure 3-3-2. Firstly, all inner voltage sources are to be short-circuit and all inner current sources are to be open-circuit. An external voltage source is connected to the two-terminal circuit and measure the current. Thus, the equivalent resistance will be calculated by

$$R_0 = \frac{U_o}{I_o} \qquad (3\text{-}3\text{-}2)$$

where U_o is the external voltage source and I_o is the current under the external voltage source condition.

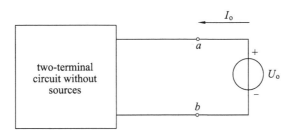

Figure 3-3-2   Voltage-current method

【Lab procedure】

(1) *Open-circuit voltage and short-circuit current*: Verify Thevenin's theorem using the circuits as shown in Figure 3-2-1. Connect the circuit as shown in Figure 3-3-3 and Figure 3-3-4, and switch on the DC power supply after checking the circuit. Measure voltage and current which are listed in Table 3-3-1. Observe the results, make records of the measured values, and fill them in Table 3-3-1.

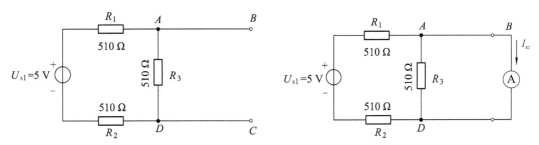

Figure 3-3-3   Open-circuit voltage   Figure 3-3-4   Short-circuit current

Table 3-3-1   Open-circuit voltage and short-circuit current

open-circuit voltage U_{AD}/V	short-circuit current I_{sc}/A	equivalent resistance/Ω

(2) *Voltage-current method*: The corresponding circuit of voltage-current method is illustrated in Figure 3-3-5. Connect the circuit as shown in Figure 3-3-5, and switch on the DC power supply after checking the circuit. Measure voltage and current which are listed in Table 3-3-2. Observe the results, make records of the measured values, and fill them in Table 3-3-2.

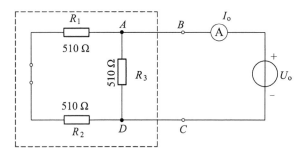

Figure 3-3-5 The voltage-current method

Table 3-3-2 The voltage-current method

voltage U_o/V	current I_o/A	equivalent resistance/Ω

【Pre-lab questions】

(1) Preview the principle of Thevenin's theorem.

(2) Preview the concept of open-circuit voltage.

(3) Preview the concept of short-circuit current.

(4) Preview the concept of equivalent resistance.

【Post-lab questions】

(1) How to verify Thevenin's theorem using the voltage and current data?

(2) What is the definition of the equivalent resistance in Thevenin's theorem?

(3) What is the definition of the open-circuit voltage?

(4) What is the definition of the short-circuit current voltage?

3.4 Impedance and power factor improvement

【Lab object】

(1) Understand the concept of impedance, reactance, and resistance.
(2) Learn to measure the impedance using a watt meter.
(3) Learn to improve the power factor methods for an inductive load.

【Lab devices】

Inductors, capacitors, transformer, watt meter, AC ammeter and voltmeter, resistors, and wires.

【Lab Principle】

Impedance is represented with the symbol Z and measured in Ohms. Impedance is expressed by

$$Z = R + jX \qquad (3\text{-}4\text{-}1)$$

where R (Resistance) is the slowing of current due to effects of the material and shape of the component. This role is the largest in resistors, but all components have at least a little resistance. X (Reactance) is the slowing of current due to electric and magnetic fields opposing changes in the current or voltage. Reactance only occurs in AC circuits and it is measured in Ohms. There are inductive reactance and capacitive reactance in a sinusoidal AC circuit.

Inductive reactance X_L is produced by inductors. These components create a magnetic field that opposes the directional changes in an AC circuit. If the current of the inductor is a sinusoidal signal, the voltage of inductor is expressed by

$$u_L = L\frac{di}{dt} = L\frac{d(I_m \sin \omega t)}{dt} = \omega L I_m \sin\left(\omega t + \frac{\pi}{2}\right) \qquad (3\text{-}4\text{-}2)$$

where I_m is the amplitude of the sinusoidal signal, ω is the angular frequency of the sinusoidal signal, and L is the inductance of the inductor. Therefore, the inductive reactance is expressed by

$$X_L = \omega L = 2\pi f L \qquad (3\text{-}4\text{-}3)$$

Capacitive reactance X_C is produced by capacitors which store an electrical charge. As current flows in an AC circuit, the capacitor charge and discharges repeatedly. If the voltage of the capacitor is a sinusoidal signal, the current is expressed by

$$i_C = C\frac{du}{dt} = C\frac{d(U_m \sin \omega t)}{dt} = \omega C U_m \sin\left(\omega t + \frac{\pi}{2}\right) \qquad (3\text{-}4\text{-}4)$$

where U_m is the amplitude of the sinusoidal signal, and C is the capacitance of the capacitor.

Therefore, the capacitive reactance is expressed by

$$X_C = \frac{1}{\omega C} = \frac{1}{2\pi fC} \quad (3\text{-}4\text{-}5)$$

Power factor is defined as the cosine of the angle between the voltage phasor and current phasor in linear AC circuits. As shown in Figure 3-4-1a, there is an inductive load in an AC voltage source. Combining the Equation 3-4-2, the phasor diagram of the inductor is illustrated in Figure 3-4-1b.

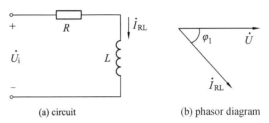

(a) circuit   (b) phasor diagram

Figure 3-4-1　Power factor of an inductive load

As shown in Figure 3-4-2a, a capacitor is connected in parallel with the inductor. Combining the Equations 3-4-2 and the Equations 3-4-4, the phasor diagram of the circuit is illustrated in Figure 3-4-2b. The angle between the voltage phasor and current phasor become smaller for inductive loads. Therefore, the power factor is raised by paralleling in a capacitor for an inductive load.

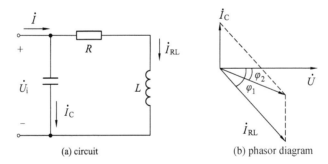

(a) circuit   (b) phasor diagram

Figure 3-4-2　The improved power factor

【Lab procedure】

(1) *AC impedance parameters*: Connect the circuit as shown in Figure 3-4-3, switch on the AC power button after checking the circuit. Make sure the RMS value of the output voltage of the transformer is below 35 V in order to prevent electric shock. Observe the results, make records of the measured values, and fill them in Table 3-4-1.

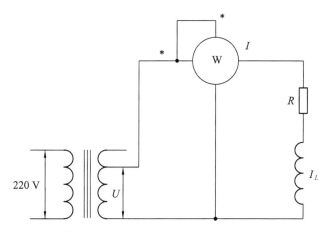

Figure 3-4-3 AC impedance parameters

Table 3-4-1 AC impedance parameters

I/A	P/W	U/V	U_1/V	U_2/V	cos φ	$|Z|/\Omega$	R/Ω	X_L/Ω	L/H

(2) *Power factor improvement*: Connect the circuit as shown in Figure 3-4-4, switch on the power button after checking the circuit. Under the condition of constant output voltage of the transformer, replace the capacitors with different capacities such as 0.47 μF, 1.0 μF, 2.2 μF, 4.3 μF, and repeat the measurement. Observe the results, make records of the measured values, and fill them in Table 3-4-2.

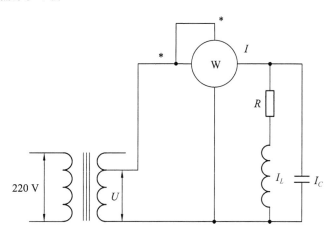

Figure 3-4-4 The power factor improvement

Table 3-4-2 The power factor improvement

No.	U/V	I/A	I_L/A	I_C/A	P/W	cos φ	C/μF
1							
2							
3							
4							

【Pre-lab questions】

(1) Preview the concept of impedance.

(2) Preview the concept of power and power factor.

(3) Preview the structure of an AC ammeter and an AC voltmeter.

【Post-lab questions】

(1) Why are the capacitors connected in parallel rather than in series, increasing the power factor for the inductive load?

(2) How to get the maximum power factor for an inductive load?

(3) What is the role of the impedance in circuits?

(4) What is the role of the power factor in circuits?

【Lab safety precautions】

(1) It is allowed to press the power button in order to prevent electric shock after the instructor confirms.

(2) Obey the wiring rules. The wires are connected in the order of ground first, then the power cord. The order of disconnecting the wires is first the power cord, then the ground.

(3) It is allowed to disconnect the wires after turning off the power.

3.5 Series RLC resonance circuits

【Lab objective】

(1) Understand the quality factor and the resonance condition in a series RLC circuit.

(2) Learn to measure the resonance frequency using the lab method.

(3) Learn to use the AC millivolt meter and oscilloscope.

【Lab devices】

Inductor, capacitors, function generator, oscilloscope, AC millivolt meter, resistors, and wires.

【Lab principle】

In a series RLC circuit shown in Figure 3-5-1, the impedance of a series RLC circuit is expressed by

$$Z = R + j\left(\omega L - \frac{1}{\omega C}\right) \tag{3-5-1}$$

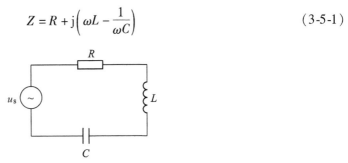

Figure 3-5-1 A series RLC circuit

A series RLC circuit may behave as either a capacitive circuit or an inductive circuit depending upon the frequency of the source voltage. However, inductive reactance will be equal to the capacitive reactance at a particular frequency. In this state, the circuit will behave like a purely resistive circuit. This phenomenon is called as resonance and the corresponding frequency is named as the resonance frequency. Resonance occurs when the imaginary part of the impedance is zero, so the following equation can be obtained.

$$\omega L = \frac{1}{\omega C} \tag{3-5-2}$$

Thus, resonance angular frequency is expressed by

$$\omega_0 = \frac{1}{\sqrt{LC}} \tag{3-5-3}$$

The quality factor is the ratio of stored power to the power consumed by the reactance and resistance of the circuit, respectively, which is expressed by

$$Q = \frac{\omega_0 L}{R} = \frac{1}{\omega_0 C R} \tag{3-5-4}$$

The Equation 3-5-4 shows that quality factor is invariable with the signal frequency. The quality factor of a resonant circuit is a measure of the "excellent" or quality of the resonant circuit. A higher value of this quality factor corresponds to a narrower bandwidth. In a series RLC circuit, current is expressed by

$$I = \frac{U_i}{|Z|} = \frac{U_i}{\sqrt{R^2 + \left(\omega L - \dfrac{1}{\omega C}\right)^2}} \tag{3-5-5}$$

As shown in Figure 3-5-2, the current is variable with the signal source frequency, which is also called frequency response. The resistance R plays an important role in signal selectivity because the rate at which the circuit current decreases as it moves away from the resonant frequency depends on the resistance R. If the resistance R is low, the current will drop sharply as it moves away from the resonant frequency. However, the resonance frequency of the circuit is not affected by the resistance R.

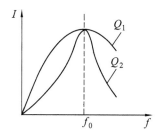

Figure 3-5-2　Frequency response

【Lab procedure】

(1) *Resonance frequency*: The circuit parameters are listed in Table 3-5-1. Connect the circuit as shown in Figure 3-5-3, switch on the AC power supply after checking the circuit. Adjust the output sinusoidal frequency of the signal generator and its amplitude is set to 1 V. There is the maximum current when the resonance occurs. In order to measure the maximum current, add a small resistor $r = 10\ \Omega$ and parallel an AC millivolt meter with the small resistor. Observe the results, make records of the measured values, and fill them in Table 3-5-1.

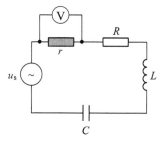

Figure 3-5-3　Resonance frequency measurement

Table 3-5-1 Resonance frequency

circuit parameters	$R = 150\ \Omega$, $L = 20$ mH, $C = 1\ \mu F$	$R = 150\ \Omega$, $L = 20$ mH, $C = 0.1\ \mu F$
measured value, f_0/Hz		
measured value, Q_0		
theoretical value, f_0/Hz		
theoretical value, Q_0		

(2) *Frequency response*: Connect the circuit as shown in Figure 3-5-3, switch on the AC power supply after checking the circuit. Adjust the output sinusoidal frequency of the signal generator and its amplitude is set to 1 V. Measure the voltage and current of the series RLC circuits. Observe the results, make records of the measured values, and fill them in Table 3-5-2.

Table 3-5-2 Amplitude-frequency response

signal frequency, f/Hz					
voltage, U_r/V					
current, I/A					

【Pre-lab questions】

(1) Preview the concept of impedance of capacitors and inductors.

(2) Preview the concept of the quality factor and frequency response.

(3) Preview the definition of initial conditions of passive elements.

【Post-lab questions】

(1) Why is an AC millivolt meter required in the lab?

(2) How to measure the quality factor in a series RLC circuit?

(3) How to analyze the resonance curve in a series RLC circuit?

(4) Why doesn't inductor allow sudden change of current?

3.6 Transient response in a first-order RC circuit

【Lab objective】

(1) Understand the zero-input response and zero-state response.

(2) Learn to identify the zero-input response and zero-state response.

(3) Learn to design differential and integrating circuits using resistors and capacitors.

【Lab devices】

Capacitors, function generator, oscilloscope, resistors, and wires.

【Lab principle】

Zero input response means that the initial conditions of the system bring a response when the input signal is zero. Zero-state response refers to the response brought by the input signal when the initial condition is zero. In a first-order RC circuit as shown in Figure 3-6-1, when the switch is turned from position "a" to position "b", the voltage across the capacitor is expressed by

$$u_c(t) = u_c(0_-) e^{-\frac{t}{\tau}} \qquad (3\text{-}6\text{-}1)$$

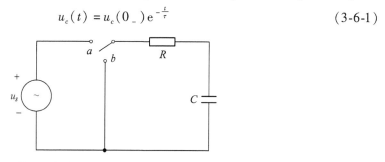

Figure 3-6-1 A first-order RC circuit

where $u_c(0_-)$ is the initial voltage of the capacitor and τ is time constant which is equal to RC. As shown in Figure 3-6-2b, the voltage of capacitor is discharge and it is also a zero-input response.

If the initial voltage of the capacitor is zero, the switch is turned from position "b" to position "a", the voltage across the capacitor is expressed by

$$u_c(t) = u_s(1 - e^{-\frac{t}{\tau}}) \qquad (3\text{-}6\text{-}2)$$

As shown in Figure 3-6-2b, the voltage of capacitor is charge by a DC voltage source and it is also a zero-state response.

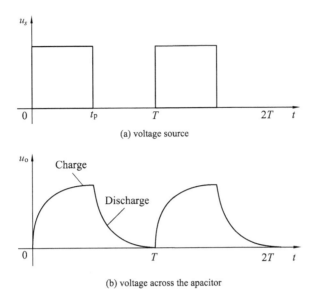

(a) voltage source

(b) voltage across the apacitor

Figure 3-6-2　Waveforms of a first-order RC circuit

When the cycle time of the input waveform is much longer than the RC time constant of the circuit, the output waveform resembles narrow positive and negative spikes as shown in Figure 3-6-3a. The positive peak of the output is generated by the rising edge of the input square wave, and the negative peak of the output is generated by the falling edge of the input square wave. Thus, it is very similar to the mathematical function of differentiation, called differential circuit as shown in Figure 3-6-4.

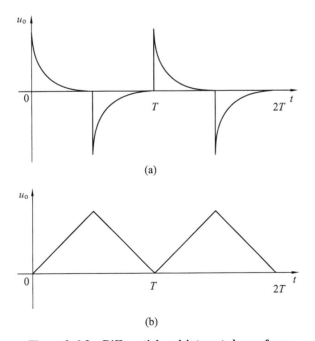

Figure 3-6-3　Differential and integrated waveform

When the cycle time of the input waveform is much smaller than the RC time constant of the circuit, the output waveform resembles triangular ramp function as shown in Figure 3-6-3b. Thus, it is very similar to the mathematical function of integration, called integral circuit as shown in Figure 3-6-5.

【Lab procedure】

(1) *Differential circuit*: The output of the signal generator is a rectangular wave with a frequency of 1 kHz. The resistance value of the resistor is 200 Ω and the capacitance value of the capacitor is 0.1 μF. Connect the circuit as shown in Figure 3-6-4, switch on the function generator after checking the circuit. Observe the output voltage waveform across the resistor using the oscilloscope. Change the different resistor, repeat the lab. Draw the output voltage waveform across the resistor.

(2) *Integral circuit*: The output of the signal generator is a rectangular wave with a frequency of 1 kHz. The resistance value of the resistor is 1 kΩ and the capacitance value of the capacitor is 0.1 μF. Connect the circuit as shown in Figure 3-6-5, switch on the function generator after checking the circuit. Change the different resistor, repeat the lab. Observe the output voltage waveform across the capacitor with the oscilloscope. Draw the output voltage waveform across the capacitor.

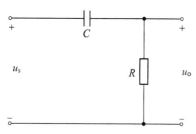
Figure 3-6-4 Differential circuit

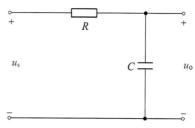
Figure 3-6-5 Integral circuit

【Pre-lab questions】

(1) Preview the concept of zero-input response and zero-state response.
(2) Preview the role of the time constant in first-order RC circuit.
(3) Preview the definition of transient response.

【Post-lab questions】

(1) How to identify the charge and discharge process of the capacitor?
(2) How to measure the time constant using the oscilloscope?
(3) Why don't capacitors allow sudden voltage changes?

3.7 Measure voltage and current in 3-phase AC circuits

【Lab object】

(1) Understand the wye and delta configuration in a 3-phase AC circuit.
(2) Understand the phase current, phase voltage, line current, and line voltage.
(3) Understand the role of the neutral wire in the 3-phase and 4-wire power system.
(4) Learn to measure the voltages and the currents in a 3-phase AC circuit.

【Lab devices】

Light bulbs, transformer, switches, AC voltmeter and ammeter, current socket and wires.

【Lab principle】

In general, transmission and distribution of electric power 3-phase system has been universally adopted because of its efficiency and simplification. Nikola Tesla (1856—1943) is best known for his contributions to the design of the modern AC electricity supply system. A supply system is symmetrical when several voltages of the same frequency have equal magnitude and are displaced from one another by equal time angle. As shown in Figure 3-7-1, phase sequence is the order or sequence in which the currents or voltages in different phases reach their maximum values one after the other. There are typical wye and delta networks in power supply systems.

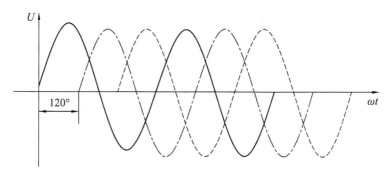

Figure 3-7-1 Waveform of 3-phase AC voltage

In a wye-connected network as shown in Figure 3-7-2, line current is equal to phase currents, and line voltage is equal to root 3 of phase voltages,

$$U_L = \sqrt{3} U_P; \quad I_L = I_P \qquad (3\text{-}7\text{-}1)$$

For balanced loads, the impedance of each phase is equal to

$$Z_U = Z_V = Z_W = Z \qquad (3\text{-}7\text{-}2)$$

Thus, the current of neutral line is equal to

$$\dot{I}_N = \dot{I}_U + \dot{I}_V + \dot{I}_W = 0 \tag{3-7-3}$$

where subscript "L" means line, and the subscript "P" means phase.

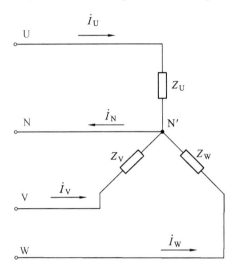

Figure 3-7-2　Wye-connected network

For balanced loads with delta-connected networks as shown in Figure 3-7-3, line voltage is equal to phase voltage, and line current is equal to root 3 of phase currents,

$$U_L = U_P;\ I_L = \sqrt{3}I_P \tag{3-7-4}$$

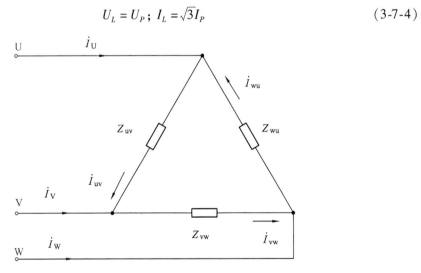

Figure 3-7-3　Delta-connected network

A 3-phase, 4-wire delta and wye network is suitable for unbalanced loads. Since the loads are unbalanced, the line currents will be different in magnitude and displaced from one another by unequal angles. The current in the neutral wire will be the phasor sum of the three-line currents. In addition, notice that single phase loads between any line and the neutral wire can be connected.

【Lab procedure】

(1) *Wye-connected networks*: Connect the circuit as shown in Figure 3-7-4, switch on the AC power supply after checking the circuit. FU means fuse elements. A wye network consisting of balanced or unbalanced loads is built by switching bulbs. Measure line current, line voltage, phase current, phase voltage and neutral current. Make records of measured value and fill them in Table 3-7-1.

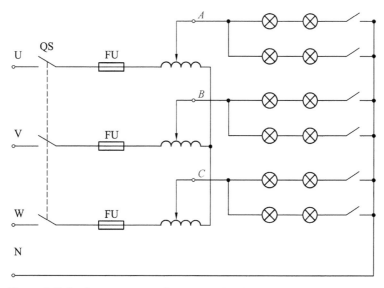

Figure 3-7-4 A wye-connected network of balanced or unbalanced loads

Table 3-7-1 A wye-connected network

items	bulbs	$I_{AL}, I_{BL}, I_{CL}/A$	$U_{AL}, U_{BL}, U_{CL}/V$	$I_{AP}, I_{BP}, I_{CP}/A$	$U_{AP}, U_{BP}, U_{CP}/V$	I_N/A
balanced load						
unbalanced load						

(2) *Delta-connected networks*: Connect the circuit as shown in Figure 3-7-5, switch on the AC power supply after checking the circuit. FU means fuse elements. A delta network consisting of balanced or unbalanced loads is built by switching bulbs. Measure line current, line voltage, phase current, and phase voltage. Make records of measured value and fill them in Table 3-7-2.

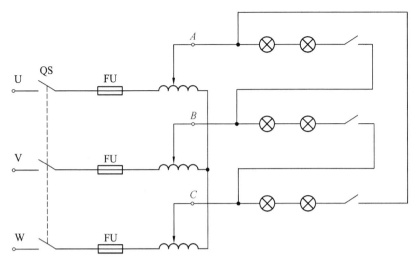

Figure 3-7-5 A delta-connected network of balanced and unbalanced load

Table 3-7-2 A delta-connected network

items	bulbs	$I_{AL}, I_{BL}, I_{CL}/A$	$U_{AL}, U_{BL}, U_{CL}/V$	$I_{AP}, I_{BP}, I_{CP}/A$
balanced load				
unbalanced load				

【Pre-lab questions】

(1) Preview the wye and the delta configurations in a 3-phase AC circuit.

(2) Preview the phase current, phase voltage, line current and line voltage.

(3) Preview the balanced loads and unbalanced loads.

【Post-lab questions】

(1) What is the role of the neutral wire in a 3-phase and 4-wires system?

(2) What is the difference between a wye and in a 3-phase AC circuit?

(3) What is the difference between the balanced loads and unbalanced loads of delta networks?

(4) What is the difference between the balanced loads and unbalanced loads of wye networks?

【Lab safety precautions】

(1) It is allowed to press the power button in order to prevent electric shock after the instructor confirms.

(2) Obey the wiring rules. The wires are connected in the order of ground first, and then the power cord. The order of disconnecting the wires is first the power cord, and then the ground.

(3) It is only allowed to disconnect the wires after turning off the power.

3.8 Power measurement in 3-phase AC circuits

【Lab object】

(1) Learn to use the watt meter in a 3-phase AC circuit.
(2) Understand the apparent power, active power, and reactive power.
(3) Learn to measure the power in a 3-phase AC circuit.

【Lab devices】

Light bulbs, transformer, switches, watt meter, AC ammeter and voltmeter, and wires.

【Lab principle】

Apparent power is a measure of alternating current power, which defines the product of root mean square value voltage and root mean square value current. The active power is actually utilized in an AC circuit. However, reactive power is the complex power in loads, such as capacitors and inductors. Reactive power represents an energy exchange between the power source and the reactive loads where no net power is gained or lost. Reactive power is stored in and discharged by inductive motors, transformers, solenoids and capacitors. The relationship between the apparent power, active power and reactive power is illustrated in Figure 3-8-1.

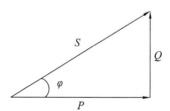

Figure 3-8-1 Power triangle

$$S^2 = P^2 + Q^2 \tag{3-8-1}$$

where S denotes the apparent power which is measured in the unit of volt-amps (VA); P denotes the active power which is measured in the unit of Watts; Q denotes the reactive power which is measured in the unit of Var, and the meaning of $\cos \varphi$ is equal to power factor.

As shown in Figure 3-8-2, regardless of whether the load is balanced or unbalanced, any phase power for a wye network can be measured with one watt meter. However, the repeat measurement is required and it is inefficiency.

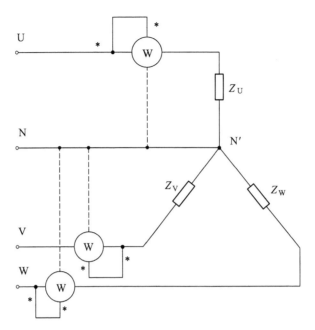

Figure 3-8-2 One watt meter method

As shown in Figure 3-8-3, regardless of whether the load is balanced or unbalanced, any phase power for a delta networks can be measured with two watt meters. Moreover, the repeat measurement isn't required and it is efficiency.

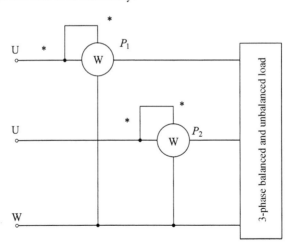

Figure 3-8-3 Two watt meters method

【Lab procedure】

(1) *One watt meter power measurement*: Connect the circuit as shown in Figure 3-8-4, switch on the AC power supply after checking the circuit. The AC voltmeter and ammeter are attached to monitor the line/phase voltages and currents for overload protection. Observe the results, make records of the measured values, and fill them in Table 3-8-1.

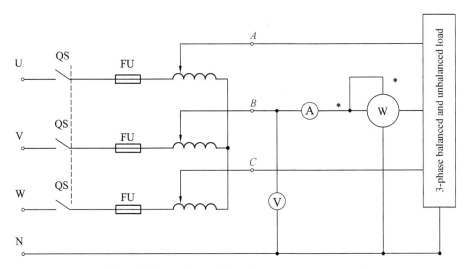

Figure 3-8-4　One watt meter power measurement

Table 3-8-1　one watt meter method

items	bulbs	P_A/W	P_B/W	P_C/W	ΣP/W
balanced load					
unbalanced load					

(2) *Two watt meters power measurement*: Connect the circuit as shown in Figure 3-8-5, switch on the AC power supply after checking the circuit. In this circuit, the AC voltmeter and ammeter are attached to monitor the line/phase voltage and current for overload protection. Observe the results, make records of the measured values, and fill them in Table 3-8-2.

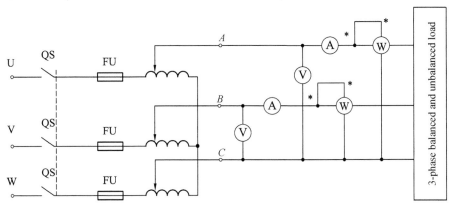

Figure 3-8-5　Two watt meters power measurement

Table 3-8-2　Two watt meters method in 3-phase and 3-wire system

items	bulbs	P_1/W	P_2/W	ΣP/W
balanced load				
unbalanced load				

【Pre-lab questions】

(1) Preview the apparent power, active power and reactive power.

(2) Preview the external wiring socket function of a watt meter.

(3) Preview the structure of a watt meter.

【Post-lab questions】

(1) What is the role of the AC ammeter and voltmeter in the power measurement?

(2) What is the cause of active power?

(3) What is the cause of reactive power?

(4) What does a watt meter measure?

【Lab safety precautions】

(1) It is allowed to press the power button in order to prevent electric shock after the instructor confirms.

(2) Obey the wiring rules. The wires are connected in the order of ground first, then the power cord. The order of disconnecting the wires is first the power cord, then the ground.

(3) It is allowed to disconnect the wires after turning off the power.

3.9 Single-phase voltage transformers

【Lab objective】

(1) Understand the nameplate and structure of single-phase voltage transformers.
(2) Learn to identify the polarity of single-phase voltage transformers.
(3) Learn to measure the external characteristics of a voltage transformer.
(4) Learn to measure the voltage transformer with no load.

【Lab devices】

Single-phase voltage transformer, variable resistors, AC voltmeter and ammeter, resistors, and wires.

【Lab principle】

Voltage transformers are capable of either increasing or decreasing the voltage and current levels of their supply, without modifying its frequency, or the amount of electrical power being transferred from one winding to another via the magnetic circuit. A single-phase voltage transformer basically consists of two electrical coils of wire, one called the "primary winding" and the other called the "secondary winding". In a single-phase voltage transformer, the primary is usually the side with the higher voltage. The turns ratio of the transformer is expressed by

$$\rho = \frac{U_1}{U_2} = \frac{N_1}{N_2} \tag{3-9-1}$$

where U_1 and U_2 are the voltage of primary and secondary, respectively; N_1 and N_2 are the numbers of primary windings and secondary windings, respectively. As shown in Figure 3-9-1a, a transformer with no load refers to that its secondary is open when its primary is connected to AC power supply. No-load losses are caused by the magnetizing current needed to energize the core of the transformer. Therefore, no-load losses will not vary according to the load on the transformer. There are mainly hysteresis losses, eddy current losses, and dielectric losses.

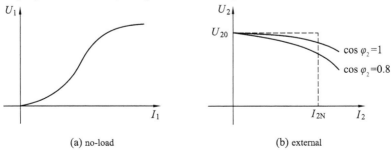

(a) no-load   (b) external

Figure 3-9-1 The transformer features

In the case where the primary voltage remains constant, the second voltage changes as the resistive load changes. This is the external feature of the transformer as shown in Figure 3-9-1b.

As shown in Figure 3-9-2, there are additive polarity and subtractive polarity to determine the polarity of a voltage transformer. In an addictive polarity, the formula is expressed by

$$U_3 = U_1 + U_2 \qquad (3\text{-}9\text{-}2)$$

where U_1 is the voltage across the primary side, U_2 is the voltage across the second side, and U_3 is the voltage between the primary side and second side.

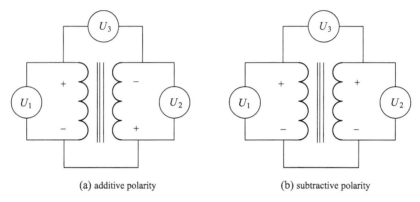

(a) additive polarity   (b) subtractive polarity

Figure 3-9-2   Identify the polarity for a voltage transformer

In a subtractive polarity, the formula is expressed by

$$U_3 = \begin{cases} U_1 - U_2 \\ U_2 - U_1 \end{cases} \qquad (3\text{-}9\text{-}3)$$

The efficiency of transformer is defined as the ratio of output power to input power. There is no loss due to friction and wind resistance because the transformer has no moving parts. As a result, its efficiency is very high and can be at least equal to 90%.

【Lab procedure】

(1) *Voltage transformer polarity and voltage ratio*: Connect the circuit as shown in Figure 3-9-3, and switch on the AC power supply after checking the circuits. Observe the results, make records of the measured values, and fill them in Table 3-9-1.

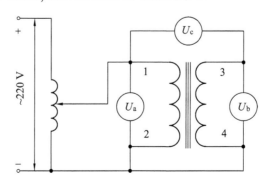

Figure 3-9-3   The polarity and voltage ratio of a voltage transformer

Table 3-9-1 Voltage transformer polarity and voltage ratio

items	U_a/V	U_b/V	U_c/V	U_a/U_b	polarity
measured value					

(2) *A voltage transformer on no load*: Connect the circuit as shown in Figure 3-9-4, and switch on the AC power supply after checking the circuits. Measure a set of voltages and currents of a voltage transformer by adjusting the output of voltage source. Observe the results, make records of the measured values, and fill them in Table 3-9-2.

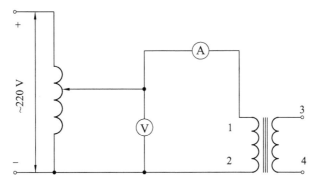

Figure 3-9-4 A voltage transformer with no load

Table 3-9-2 A voltage transformer on no load

voltage/V	30	60	90	120	150
current/A					

(3) *External characteristics of a voltage transformer*: Connect the circuit as shown in Figure 3-9-5, and switch on the AC power supply after checking the circuit. When the voltage on the primary side is 36 V, a set of voltages and currents on the second side of the voltage transformer are measured by adjusting the variable resistor. Observe the results, make records of the measured values, and fill them in the Table 3-9-3.

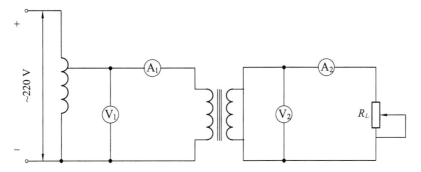

Figure 3-9-5 External characteristics of a voltage transformer

Table 3-9-3 External characteristics of a voltage transformer

R_L/Ω					
U_{second}/V					
I_{second}/A					

【Pre-lab questions】

(1) Preview the structure of a single-phase voltage transformer.

(2) Preview the nameplate of a single-phase voltage transformer.

(3) Preview the polarity and characteristics of a single-phase voltage transformer.

【Post-lab questions】

(1) Why does the voltage of the primacy side keep constants when external characteristics of a single-phase voltage transformer is measured?

(2) What is the definition of the polarity of a single-phase voltage transformer?

(3) What is the definition of the ratio of a single-phase voltage transformer?

(4) Why is the efficiency of transformer higher than rotating machines?

【Lab safety precautions】

(1) It is allowed to press the power button in order to prevent electric shock after the instructor confirms.

(2) Obey the wiring rules. The wires are connected in the order of ground first, then the power cord. The order of disconnecting the wires is first the power cord, then the ground.

(3) It is allowed to disconnect the wires after turning off the power.

3.10 Start 3-phase induction motors

【Lab objective】

(1) Understand the nameplate and structure of squirrel cage motors.

(2) Understand the starting methods of squirrel cage motors.

(3) Learn to operate the starting methods of squirrel cage motors.

【Lab devices】

Squirrel cage motor, normally closed buttons, normally open buttons, coils, switches, contacts, fuse elements, thermal overload relay, AC ammeter, and wires.

【Lab principle】

An electrical motor is an electro-mechanical device which converts electrical energy into mechanical energy. A 3-phase induction motor consists of a stator and a rotor. As shown in Figure 3-10-1, the stator is made up of numbers of slots which can construct a 3-phase winding circuit and be wired with 3-phase wye-mode or delta-mode AC source. The 3-phase winding is arranged in such a manner in the slots that can produce one rotating magnetic field. The rotor includes a cylindrical laminated core with parallel slots that can carry conductors. The conductors are heavy copper or aluminum bars fitted in each slot and short-circuited by the end rings. The slots are not exactly made parallel to the axis of the shaft but are slotted a little skewed, because this arrangement reduces magnetic humming noise and can avoid stalling of the motor.

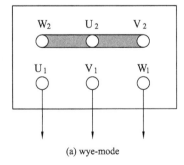 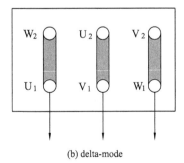

(a) wye-mode   (b) delta-mode

Figure 3-10-1 Stator winding wires

The most widely used motor is the squirrel cage induction motor. The starting methods of a squirrel cage induction motor are classified into four methods: (i) direct-on-line starter (DOL); (ii) star delta starter; (iii) auto-transformer starter; (iv) rotor resistance control of induction motor.

As shown in Figure 3-10-2, direct-on-line starter is connected directly between the power supply and motor terminal. "M3 ~" means 3-phase induction motor, "S" means the main circuit switch which serves as the main power supply switch, "FU" means fuse elements, and "FR" means thermal overload relay. "FU" and "FR" components are safety protection mechanism for motor. The motor at the time of starting draws very high starting current (about 5 to 7 times of the full load current) for the very short duration. The amount of current drawn by the motor depends upon its design and size. However, such a high value of current does not harm the motor because benefited from rugged structure of the squirrel cage induction motor. Compared to the other methods, direct-on-line starter is the simplest and inexpensive method for starting induction motor.

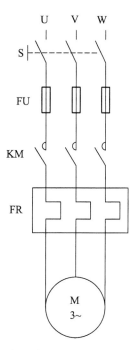

Figure 3-10-2　Direct-on-line starter

In order to reduce the motor starting current and the disturbances from the electrical supply, large induction motors are started at reduced voltage and then have full supply voltage. As shown in Figure 3-10-3, voltage reduction during star-delta starting is achieved by physically reconfiguring the motor windings. There are open type transition and closed type transition during the transition from star mode to delta mode.

In open type transition there are four states:

(1) OFF state: All contactors are open.

(2) Star state: The main (KM_1) and the star (KM_3) contactors are closed, the delta (KM_2) contactor is open. The motor is connected in star and will produce one third of DOL torque at one third of DOL current.

(3) Open state: The main contractor is closed, the delta and star contactors are open.

There is voltage on one end of the motor windings, but the other end is open, so no current can flow. The motor has a spinning rotor and behaves like a generator.

(4) Delta state: The main and the delta contactors are closed. The star contactor is open. The motor is connected to full line voltage and full power, and torque are available.

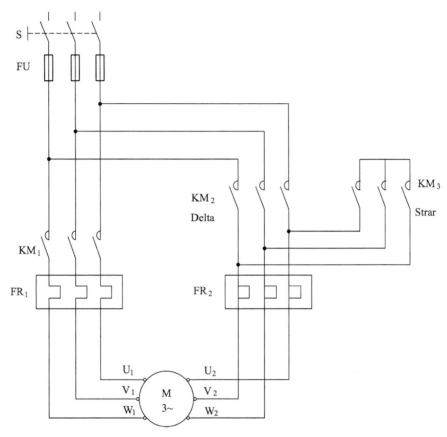

Figure 3-10-3 Star-delta starter

In closed type transition there are five states:

(1) OFF state: All contactors are open.

(2) Star state: The main (KM_1) and the star (KM_3) contactors are closed and the delta (KM_2) contactor is open. The motor is connected in star and will produce one third of DOL torque at one third of DOL current.

(3) Star transition state: The motor is connected in star and the resistors are connected across the delta contactor via the aux contactor.

(4) Closed transition state: The main (KM_1) contactor is closed and the delta (KM_2) and star (KM_3) contactors are open. Current flows through the motor windings and the transition resistors via aux contactor.

(5) Delta state: The main and the delta contactors are closed. The transition resistors are shorted out. The star contactor is open. The motor is connected to full line voltage and full power, and torque are available.

The power is disconnected from the motor while the windings are reconfigured via external switching in open type transition. However, the power is maintained to the motor at all time in closed transition. The contacts are controlled by the timer built into the starter. The star and delta are electrically interlocked and preferably mechanically interlocked as well.

【Lab procedure】

Before the lab, read the nameplate of the squirrel cage motor carefully and observe the connection mode of six terminals in the junction box.

(1) *Direct-on-line starter*: Connect the circuit as shown in Figure 3-10-2, switch on the AC power supply after checking the circuit. Observe the starting current, make records of measured value.

(2) *Star-delta starter*: Connect the circuit as shown in Figure 3-10-3, switch on the AC power supply after checking the circuit. The transition from star mode to delta mode is controlled by the timer. This process is done automatically. Observe the starting current of the star-delta starter. Compare starting current between direct-on-line method and star-delta method.

【Pre-lab questions】

(1) Preview the nameplate and structure of 3-phase induction motors.

(2) Preview the starting methods of 3-phase induction motors.

【Post-lab questions】

(1) What are the results if the connect wires are mistaken between star mode and delta modes?

(2) What is the rated current difference between star mode and delta mode of an induction motor?

【Lab safety precautions】

(1) It is allowed to press the power button in order to prevent electric shock after the instructor confirms.

(2) Obey the wiring rules. The wires are connected in the order of ground first, then the power cord. The order of disconnecting the wires is first the power cord, then the ground.

(3) It is allowed to disconnect the wires after turning off the power.

Chapter 3 Electrical Engineering Lab Parts

3.11 Jog and continuous control of motors

【Lab objective】

(1) Understand the principle of jog and continuous control of 3-phase induction motors.

(2) Understand the principle of self-locking and interlock in the 3-phase induction motors.

(3) Learn to operate the jog and continue rotation of 3-phase induction motors.

【Lab devices】

Squirrel cage motor, normally closed buttons, normally open buttons, coils, switches, contacts, fuse elements, thermal overload relay, AC ammeter, and wires.

【Lab principle】

In a 3-phase induction motor, the jog control is important to create a circuit that will allow the operator to momentarily energize the circuit without pressing the stop button. In a jog control circuit as shown in Figure 3-11-1, there are stop button (SB_2), start button (SB_1) and the contact coil (KM). When the switch is closed and the start button is pressed, the control circuit is to energize the coil KM and the main contact (KM) is closed, the motor starts to rotate. When the start button is released, the control circuit will be de-energized by the coil KM and the main contact (KM) will be cut off, the motor will stop. There is no need to press the stop button (SB_2) during the jog control process of a 3-phase induction motor.

Compared to the jog control, add a normally closed auxiliary contact KM_2 in a continuous control circuit as shown in Figure 3-11-2. When the switch is closed and SB_1 button is pressed, the control circuit is to energize the coil KM and the main contact (KM) and the normally open auxiliary contact KM_2 are closed, the motor starts to rotate. Next, when the SB_1 button is released, the control circuit is also connected to energize the main contact (KM) by the normally closed auxiliary contact KM_2, the motor continues to rotate. When the SB_2 button is pressed, the control circuit is cut off and the motor stops.

The Lab Tutorial of Electrical Engineering and Electronics

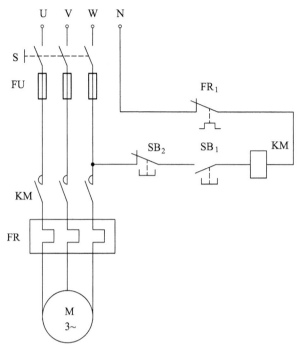

Figure 3-11-1 The jog control of a 3-phase induction motor

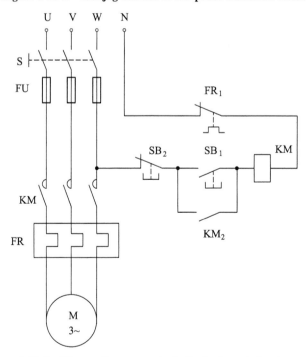

Figure 3-11-2 The continuous control of a 3-phase induction motor

Both jog control and continuous control of 3-phase induction motor are illustrated in Figure 3-11-3. The SB_1 button is to run the motors at continuous control mode, the SB_2 button is to stop the motors at the continuous control mode, and the SB_3 button is to run the motors at the jog control mode.

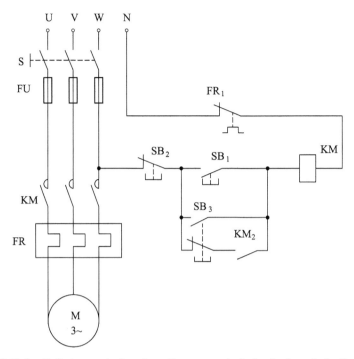

Figure 3-11-3 Both jog control and continuous control of a 3-phase induction motor

【Lab procedure】

Before the lab, read the nameplate of the squirrel cage motor carefully and observe the six terminals connected method in the junction box.

(1) *Jog control*: Connect the circuit as shown in Figure 3-11-1, switch on the AC power supply after checking the circuit.

① Press the start button (SB_1), the stator is energized and the rotor begins to rotate.

② Release the start button (SB_1), the stator is de-energized and the rotor stops.

③ Cut off the switch.

(2) *Continuous control*: Connect the circuit as shown in Figure 3-11-2, switch on the AC power supply after checking the circuit.

① Press the start button (SB_1), the stator is energized and the rotor begins to rotate.

② Release the start button (SB_1), the stator is still energized and the rotor continues to rotate.

③ Press the stop button (SB_2), the stator is de-energized and the rotor stops.

④ Cut off the switch.

(3) *Both jog and Continuous control*: Connect the circuit as shown in Figure 3-11-3, switch on the AC power supply after checking the circuit.

① *Jog control*: Press the start button (SB_3), the stator of motor is energized and the rotor starts to rotate. Release the start button (SB_3), the stator of motor is de-energized and the rotor stops. Cut off the switch.

② *Continuous control*: Press the start button (SB_1), the stator of motor is energized and the rotor starts to rotate. Release the start button (SB_1), the stator of motor is still energized and the rotor continues to rotate. Press the stop button (SB_2), the stator of motor is de-energized and the rotor stops. Cut off the switch.

【Pre-lab questions】

(1) Preview the principle of the jog control of 3-phase induction motors.

(2) Preview the principle of the continuous control of 3-phase induction motors.

(3) Preview the function of normally open button and normally closed button.

【Pro-lab questions】

(1) What is the role of the thermal overload relay?

(2) What is the role of the contacts?

(3) What is the difference between the normally closed buttons and normally open buttons?

【Lab safety precautions】

(1) It is allowed to press the power button in order to prevent electric shock after the instructor confirms.

(2) Obey the wire rules. The wires are connected in the order of ground first, then the power cord. The order of disconnecting the wires is first the power cord, then the ground.

(3) It is allowed to disconnect the wires after turning off the power.

3.12 Forward and reverse rotation of motors

【Lab objective】

(1) Understand the principle of forward direction of 3-phase induction motors.
(2) Understand the principle of reverse direction of 3-phase induction motors.
(3) Learn to operate the forward and reverse rotation of 3-phase induction motors.

【Lab devices】

Squirrel cage motor, normally closed buttons, normally open buttons, coils, switches, contacts, fuse elements, thermal overload relay, AC ammeter, and wires.

【Lab principle】

To change the direction of a 3-phase induction machine rotation, two of its phases need to be exchanged. This process can be implemented by using two contacts, one for the forward rotation and the other for the reverse rotation. The forward and reverse contacts are required to interlock, which means one contact should be de-energized before the other one can be energized. There are mechanical interlocking and electrical interlocking. Mechanical interlocking uses the contacts to operate a mechanical lever that prevents the other contact from closing while one is energized. Electrical interlocking uses double acting push buttons and auxiliary contacts.

The forward and reverse direction control of 3-phase induction motor is illustrated in Figure 3-12-1. The role of button KM_{12} and KM_{22} is forward and reverse direction of self-locking, separately. The normally closed part of the SB_1 button is connected in series with KM_2 coil, and the normally closed part of the SB_2 button is connected in series with KM_1 coil.

As shown in Figure 3-12-1, the dashed lines drawn between the push buttons indicate that they are mechanically connected in Figure 3-12-1. Both push buttons will push at the same time, such as SB_1 button and R button, SB_2 button and F button. When the SB_1 button is pressed, the motor is running in the forward direction. The normally closed part of the SB_1 button will open and disconnect KM_2 coil from the line before the normally open part closes to energize KM_1 coil. When the SB_2 button is pressed, the motor is running in the reverse direction. The normally closed part of the SB_2 button will open and disconnect KM_1 coil from the line before the normally open part closes to energize KM_2 coil.

The Lab Tutorial of Electrical Engineering and Electronics

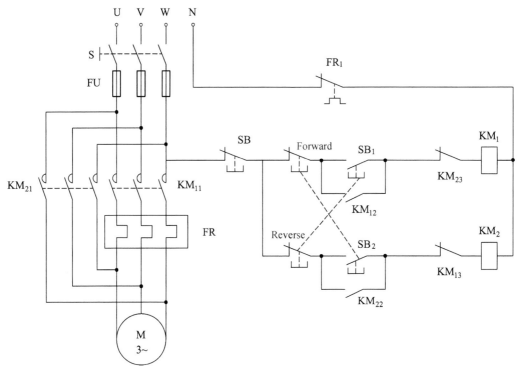

Figure 3-12-1   The forward and reverse direction control of 3-phase induction motors

As shown in Figure 3-12-1, the electrical interlocking is accomplished by the normally closed auxiliary contacts on one contact KM_{12} in series with the coil of the other contact KM_{23} (Similarly, KM_{22} in series with KM_{13}). The SB_1 push button is pressed and KM_1 coil energizes. This causes KM_{11}, KM_{12} and KM_{13} contacts to change position. The KM_{11} load contacts close and connect the motor to the line. The normally open KM_{12} auxiliary contact closes to maintain the circuit when the SB_1 push button is released, and the normally closed KM_{13} auxiliary contact connected in series with KM_2 coil opens.

【Lab procedure】

Connect the circuit as shown in Figure 3-12-1, switch on the AC power supply after checking the circuit.

(1) *Forward direction rotation*: Press the start button (SB_1), the main contact coil (KM_1) is energized, current is flowing through the stator and the rotor begins to rotate forwardly.

(2) *Reverse direction rotation*: Press the start button (SB_2), the main contact coil (KM_2) is energized, current is flowing through the stator and the rotor begins to rotate reversely.

(3) Press the stop button and cut off the switch.

(4) Notice the motor can be rotated from forward direction to reverse direction without pressing the stop button.

【Pre-lab questions】

(1) Preview the principle of the forward direction of 3-phase induction motors.

(2) Preview the principle of the reverse direction of 3-phase induction motors.

(3) Preview the mechanical interlocking and electrical interlocking.

【Pro-lab questions】

(1) What is the interlock circuit in the 3-phase induction motors?

(2) How to identify the forward direction of 3-phase induction motors?

(3) How to identify the reverse direction of 3-phase induction motors?

【Lab safety precautions】

(1) It is allowed to press the power button in order to prevent electric shock after the instructor confirms.

(2) Obey the wiring rules. The wires are connected in the order of ground first, then the power cord. The order of disconnecting the wires is first the power cord, then the ground.

(3) It is allowed to disconnect the wires after turning off the power.

3.13 Test dependent sources

【Lab objective】

(1) Understand the principle of the dependent sources.

(2) Understand the relationship between operational amplifiers and a dependent source.

(3) Learn to analyze the circuits including the dependent sources.

【Lab devices】

Adjustable DC regulated power supply, op amp, DC ammeter and voltmeter, resistors, and wires.

【Lab principle】

The source which supplies the active power to the network is known as the electrical source. There are independent source and dependent source, also called controlled source. The output voltage or current of dependent source is not fixed but depends on the voltage or current in another part of the circuit. There are four types of dependent source, namely voltage controlled voltage source (VCVS), voltage controlled current source (VCCS), current controlled voltage source (CCVS), and current controlled current source (CCCS).

As shown in Figure 3-13-1, the VCVS circuit is accomplished by an operational amplifier. Due to the virtual short and virtual open characteristic of op amp, the formula is expressed by

$$u_+ = u_- = u_1, \quad i_{R_1} = i_{R_2} \tag{3-13-1}$$

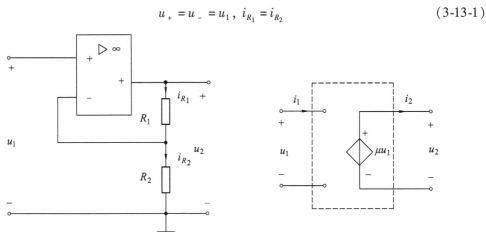

(a) op amp (b) circuit model

Figure 3-13-1 The VCVS circuit

The output voltage of op amp is expressed by

$$u_2 = i_{R_1}R_1 + i_{R_2}R_2 = i_{R_2}(R_1 + R_2)$$
$$= \frac{u_1}{R_2}(R_1 + R_2) = \left(1 + \frac{R_1}{R_2}\right)u_1$$
$$= \mu u_1 \tag{3-13-2}$$

where μ is the proportionality constants or gains. It is the ratio of the voltage output to the voltage input.

As shown in Figure 3-13-2, the VCCS circuit is accomplished by an operational amplifier. The output current of op amp is expressed by

$$i_2 = i_R = \frac{u_+}{R_2} = \frac{u_-}{R_2} = \frac{u_1}{R_2} \tag{3-13-3}$$

$$g = \frac{i_2}{u_1} = \frac{1}{R_2} \tag{3-13-4}$$

where g is also proportionality constant. It is called the transconductance.

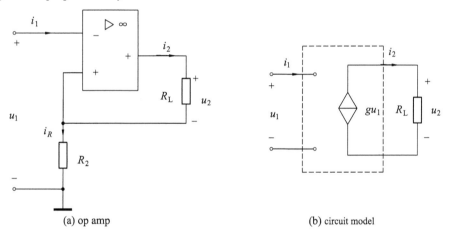

(a) op amp (b) circuit model

Figure 3-13-2 The VCCS circuit

As shown in Figure 3-13-3, the CCVS circuit is accomplished by an operational amplifier. Due to the virtual open characteristic of op amp, the output voltage of op amp is expressed by

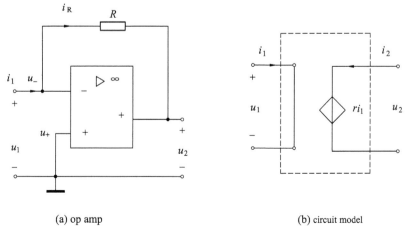

(a) op amp (b) circuit model

Figure 3-13-3 The CCVS circuit

$$u_2 = -i_1 R \tag{3-13-5}$$

$$r = \frac{u_2}{i_1} = -R \tag{3-13-6}$$

where r is also the proportionality constant. It is called the trans-resistance because it takes the form of Ohm's law.

As shown in Figure 3-13-4, the CCCS circuit is accomplished by an operational amplifier. The output voltage of op amp is expressed by

$$-i_{R_1} R_1 = i_1 R_1, \quad i_{R_2} = -\frac{i_1 R_1}{R_2} \tag{3-13-7}$$

$$i_2 = -(i_{R_1} + i_{R_2}) = i_1 + \frac{i_1 R_1}{R_2} = \left(1 + \frac{R_1}{R_2}\right) i_1$$

$$\alpha = \frac{i_2}{i_1} = 1 + \frac{R_1}{R_2} \tag{3-13-8}$$

where α is also the proportionality constant. It is called as the current gain because it is the ratio of current output to current input.

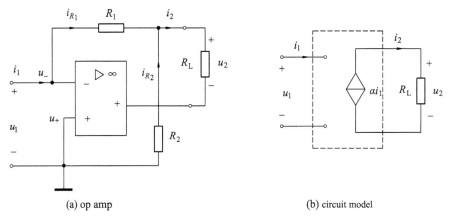

Figure 3-13-4 The CCCS circuit

【Lab procedure】

(1) VCVS: Connect the circuit as shown in Figure 3-13-5, switch on the DC power after checking the circuit. Adjust the DC voltage source and the variable resistors which are listed in Table 3-13-1. Observe the results, make records of the measured values, and fill them in the Table 3-13-1.

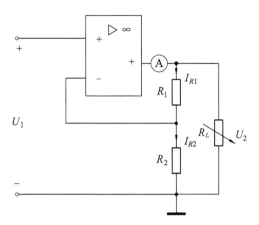

Figure 3-13-5 The VCVS test circuit

Table 3-13-1 The VCVS test

items	$R_L(U_1 = 3\ \text{V})$			$U_1(R_L = 2\ \text{k}\Omega)$		
	1 kΩ	2 kΩ	3 kΩ	1 V	2 V	3 V
U_2						
measured value μ						
theoretical value μ						

(2) *VCCS*: Connect the circuit as shown in Figure 3-13-6, switch on the DC power after checking the circuit. Adjust the DC voltage source and the variable resistors which are listed in Table 3-13-2. Observe the results, make records of the measured values, and fill them in the Table 3-13-2.

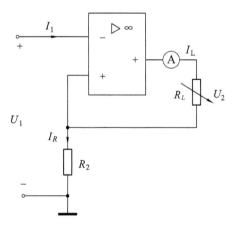

Figure 3-13-6 VCCS test circuits

Table 3-13-2 VCCS test

items	$R_L(U_1 = 3\text{ V})$			$U_1(R_L = 2\text{ k}\Omega)$		
	1 kΩ	2 kΩ	3 kΩ	1 V	2 V	3 V
I_L						
measured value g						
theoretical value g						

(3) *CCVS*: Connect the circuit as shown in Figure 3-13-7, switch on the DC power after checking the circuit. Adjust the DC voltage source and the variable resistors which are listed in Table 3-13-3. Observe the results, make records of the measured values, and fill them in Table 3-13-3.

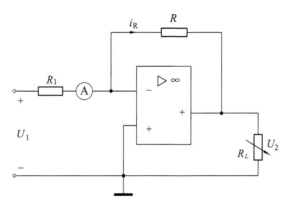

Figure 3-13-7 CCVS test circuit

Table 3-13-3 CCVS test

items	$R_L(U_1 = 3\text{ V})$			$U_1(R_L = 2\text{ k}\Omega)$		
	1 kΩ	2 kΩ	3 kΩ	1 V	2 V	3 V
I_1						
U_2						
measured value r						
theoretical value r						

(4) *CCCS*: Connect the circuit as shown in Figure 3-13-8, switch on the DC power after checking the circuit. Adjust the DC voltage source and the variable resistors which are listed in Table 3-13-4. Observe the results, make records of the measured values, and fill them in Table 3-13-4.

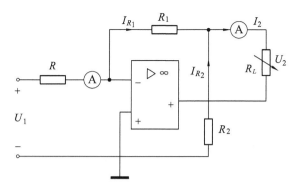

Figure 3-13-8 CCCS test circuit

Table 3-13-4 CCCS test

items	$R_L (U_1 = 3\ \text{V})$			$U_1 (R_L = 2\ \text{k}\Omega)$		
	1 kΩ	2 kΩ	3 kΩ	1 V	2 V	3 V
I_1						
I_2						
measured value α						
theoretical value α						

【Pre-lab questions】

(1) Preview the principle of the dependent source VCVS and VCCS.

(2) Preview the principle of the dependent source CCVS and CCCS.

(3) Preview the principle and function of operational amplifier.

【Pro-lab questions】

(1) What is the difference between an independent source and a dependent source?

(2) Why is an additional DC voltage source connected with the op amp?

(3) Is the output characteristic of a dependent source suitable for AC signals?

3.14 The response of second order circuits

【Lab objective】

(1) Understand the overdamped and underdamped cases in second order circuits.

(2) Understand the critically damped case in second order circuits.

(3) Understand the status trajectory of second order circuits.

(4) Learn to measure the natural frequency and damped factor of the underdamped case.

【Lab devices】

Function generator, oscilloscope, capacitors, inductors, resistors, and wires.

【Lab principle】

As shown in Figure 3-14-1, a second-order circuit is characterized by a second-order differential equation, which consists of resistors and the equivalent of two energy storage elements such as inductor and capacitor. At the initial time $t(0_-)$, the switch is at the position "a", the initial voltage of capacitor is denoted "$u_C(0_-)$" and the initial current of inductor is denoted "$i_L(0_-)$". When the switch is moved from the position "a" to the position "b", apply the KVL with the loop, the current of a series RLC circuit is expressed by

$$Ri + L\frac{di}{dt} + \frac{1}{C}\int_{-\infty}^{t} i\,dt = 0 \tag{3-14-1}$$

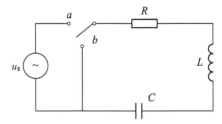

Figure 3-14-1 A second order circuit

Do differential to the Equation 3-14-1, we get

$$\frac{d^2 i}{dt^2} + \frac{R}{L}\frac{di}{dt} + \frac{i}{CL} = 0 \tag{3-14-2}$$

The characteristic root of the Equation 3-14-2 is equal to

$$S_{1,2} = -\frac{R}{2L} \pm \sqrt{\left(\frac{R}{2L}\right)^2 - \frac{1}{CL}} \tag{3-14-3}$$

A more compact way of the Equation 3-14-3 is expressed by

$$S_{1,2} = -\alpha \pm \sqrt{\alpha^2 - \omega_0^2} \quad (3\text{-}14\text{-}4)$$

$$\alpha = \frac{R}{2L}, \ \omega_0 = \frac{1}{\sqrt{CL}} \quad (3\text{-}14\text{-}5)$$

There are three cases in the equation 3-14-4. i. If $\alpha > \omega_0$, it is overdamped case; ii. If $\alpha = \omega_0$, it is critically damped case; iii. If $\alpha < \omega_0$, it is underdamped case. Damping is the gradual loss of the initial stored energy. The damping role is due to the presence of resistor. Oscillatory role is due to the presence of the inductor and capacitor. The critically damped case is the borderline between the underdamped and overdamped cases. In general, it is difficult to tell the difference between an overdamped response and a critically damped response from a waveform.

Considering underdamped case, the natural response is equal to

$$i(t) = e^{\alpha t}(B_1 \cos \omega_d t + jB_2 \cos \omega_d t) \quad (3\text{-}14\text{-}6)$$

where α is damping coefficient and ω_d is the resonant frequency. As shown in Figure 3-14-2, the natural response for the underdamped case is exponentially damped and oscillatory. Therefore, the parameters α and ω_d of underdamped case can be derived from the natural response,

$$t_2 - t_1 = T_d, \ T_d = \frac{2\pi}{\omega_d}, \ \alpha = \frac{1}{T_d}\ln\frac{i_{1m}}{i_{2m}} \quad (3\text{-}14\text{-}7)$$

If the voltage and current of the capacitor are plotted in a rectangular coordinate system, the curve is called status trajectory.

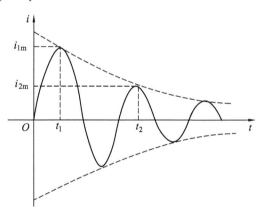

Figure 3-14-2 The natural response of underdamped case

【Lab procedure】

(1) *3 types of cases*: Connect the circuit as shown in Figure 3-14-1, and switch on the power supply after checking the circuit. The input voltage signal is from the function generator created a square wave. The output signal is the current of series RLC which is observed by oscilloscope. By adjusting the resistance value, the circuit is placed in an overdamped case, a critical damped case, and an underdamped case, respectively. Make records and draw the curve on the graph sheet.

(2) Consider the underdamped case, calculate the damped coefficient α and resonant

frequency ω_d, and fill them in Table 3-14-1.

Table 3-14-1   The underdamped case

items	ω_d	i_{1m}	i_{2m}	α	T_d
measured value					
theoretical value					

(3) *Status trajectory*: Set the oscilloscope working mode to XY, the channel 2 is connected to the current and the channel 1 is connected to the voltage across the capacitor. Consider the underdamped case and the overdamped case, observed the status trajectory. Make records and draw the curve on the graph sheet.

【Pre-lab questions】

(1) Preview the overdamped case and underdamped case in second order circuits.

(2) Preview the critically damped case in second order circuits.

(3) Preview the status trajectory of second order circuits.

【Pro-lab questions】

(1) What are the characteristics of the overdamped case, critically damped case and underdamped case in second order circuits, respectively?

(2) What is the role of the resistor in the second order circuits?

(3) How to distinguish the overdamped case and the underdamped case?

3.15 Negative impedance converters

【Lab objective】

(1) Understand the principle of the negative impedance converter.
(2) Learn to design a negative impedance converter using op amps.
(3) Learn to measure the voltage-current characteristic of the negative impedance converter.

【Lab devices】

Function generator, oscilloscope, adjustable DC regulated power supply, op amp, resistors, and wires.

【Lab principle】

Compared to a normal load that consumes energy from them, the negative impedance converter (NIC) injects energy into circuits, which is accomplished with a one-port op-amp circuit. The NIC can be divided into a negative impedance converter with voltage inversion (VNIC) and a negative impedance converter with current inversion (CNIC). The basic circuit of a CNIC is illustrated in Figure 3-15-1.

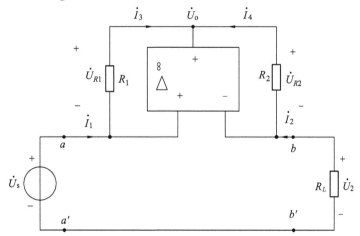

Figure 3-15-1 A negative impedance converter with current inversion

Due to virtual short characteristic of op amp, the voltage is equal to

$$\dot{U}_+ = \dot{U}_- \Rightarrow \dot{U}_1 = \dot{U}_2 \tag{3-15-1}$$

The output voltage of the op amp is

$$\dot{U}_o = \dot{U}_1 - \dot{I}_3 R_1 = \dot{U}_2 - \dot{I}_4 R_2 \tag{3-15-2}$$

Due to virtual open characteristic of op amp, the current is equal to

$$\dot{I}_1 = \dot{I}_3, \dot{I}_2 = \dot{I}_4 \qquad (3\text{-}15\text{-}3)$$

Combining with the Equations 3-15-2 and the Equation 3-15-3, it is derived by

$$\dot{I}_1 R_1 = \dot{I}_2 R_2 = -\frac{\dot{U}_1}{Z_L} R_2 \qquad (3\text{-}15\text{-}4)$$

From the perspective of the input port, the input impedance is equal to

$$Z_i = \frac{\dot{U}_1}{\dot{I}_1} = -\frac{R_1}{R_2} Z_L = -\frac{1}{K} Z_L \qquad (3\text{-}15\text{-}5)$$

where K is equal to R_2/R_1. Thus, it is equivalent to negative impedance element.

For the internal resistance case, the circuit of the negative impedance converter is illustrated in Figure 3-15-2. The external voltage and current of the a-a' port is

$$U_1 = U_s - I_1 R_s \qquad (3\text{-}15\text{-}6)$$

The external voltage and current of the b-b' port is

$$U_2 = U_s - KI_2 R_2 \qquad (3\text{-}15\text{-}7)$$

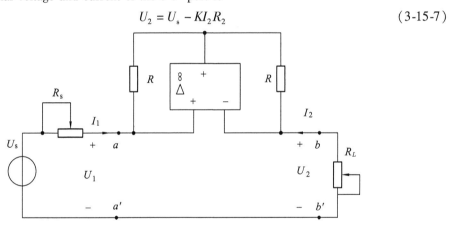

Figure 3-15-2 NIC with internal resistance

【Lab procedure】

(1) *Voltage-current characteristic*: Connect the circuit as shown in Figure 3-15-2. Check the polarity of the op amp voltage source. Turn on the power and measure the voltage and current which are listed in Table 3-15-1. Make records of the measured values, and fill them in Table 3-15-1. Draw the voltage-current curve on the graph sheet.

Table 3-15-1 Voltage-current characteristic

U_1/V	3	2	1	−1	−2	−3
U_R/V						
I_1/A						
R_L/Ω						

(2) *Phase of voltage and current*: The input voltage is the sinusoidal signal from the function generator. The voltage and current across the NIC is observed by the oscilloscope. Draw the waveform on the graph sheet.

【Pre-lab questions】

(1) Preview the principle and function of the negative impedance converters.

(2) Preview the principle and function of the operational amplifier.

【Pro-lab questions】

(1) What is the role of op amp in the negative impedance converter?

(2) Does the negative resistance converter absorb or emit power? Where is its energy from?

(3) How many types of negative impedance converters?

3.16 PLC program for star-delta motor starter

【Lab objective】

(1) Understand the basic instructions such as SET, RESET and TIMER of programmable logic controllers.

(2) Learn to properly connect wiring of motor control circuit and program debugging.

(3) Learn to design a simple program based on PLC.

【Lab devices】

Squirrel cage motor, PLC, switch, computer, fuse elements, contact, thermal overload relay, normally closed buttons, normally open buttons, coil and wires.

【Lab principle】

Programmable logic controllers (PLC) are the major components in industrial automation and control systems. A PLC program consists of a set of instructions either in textual or graphical form, which represents the logic to be implemented for specific industrial real time applications. The basic steps to design program based on PLC are briefly given as follows.

(1) Define the control task. The control task specifies what needs to be done and is defined by those who are involved in the operation of the machine or process.

(2) Define the inputs and outputs. List the input variables and output variables according to control needs.

(3) Develop a logical sequence of operation. This can be done with a flow chart or a sequence table. List all the conditions and make the design using flowchart.

(4) Develop the PLC program. Program coding is the process of translating a logic or relay diagram into PLC ladder program form. Open and configure the PLC programming software. Add the required rungs and address them. Check the errors and simulate it.

(5) Test the program. Properly connect wiring of control circuit and program debugging.

In the case of star-delta starter circuits of the 3-phase induction the motor, refer to the Figure 3-10-3. Firstly, the motor is started in the star connection and when the motor takes up speed of about (75% –80%) speed, it changes in the delta connection. The starting voltage is reduced to root 3. The connected wiring method of star-delta starter based on PLC is illustrated in Figure 3-16-1.

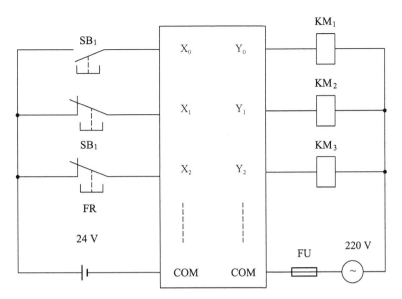

Figure 3-16-1 The connected wiring method of PLC

As shown in Figure 3-16-2, the steps of designing ladder diagram of star-delta starter are briefly given as follows.

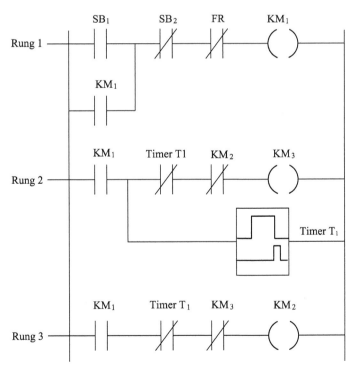

Figure 3-16-2 The ladder diagram of star-delta starter

(1) *Rung 1 main contact*: The main contact depends upon the normally open input start push button (SB_1), normally closed stop button (SB_2) and normally closed overload relay (FR). When start button is pressed while the stop button is not pressed and overload relay is not

activated, the main contact is only be energized. A normally open input named (KM_1) is added in parallel to the start button SB_1. By doing so, a push button is created which means that once motor is started, it will be kept started even if the start button is released.

(2) *Rung 2 star contact*: Star contact depends upon main contact, normally closed contacts of timer (T1), and normally closed contacts of output delta contact (KM_2). Therefore, star contact will only be energized if the main contact is ON, the time output is not activated and the delta contact is not energized. Timer T_1 measures the time after which the winding connection of star-delta starter is to be changed. It will start counting time after the main contact is energized.

(3) *Rung 3 delta contact*: Delta contact will be energized when the main contact (KM_1) is energized, timer T_1 is activated and the star contact (KM_3) is de-energized.

【Lab procedure】

Connected the circuit as shown in Figure 3-10-3 and Figure 3-16-1. Simulate the ladder diagram of star-delta starter as shown in Figure 3-16-2. Observe the results of star and delta starter of the 3-phase induction motor.

【Pre-lab questions】

(1) Preview the basic instructions such as SET, RESET and TIMER of programmable logic controllers.

(2) Preview the wye starter and delta starter of 3-phase induction motor.

【Pro-lab questions】

(1) What is the main difference between PLC control system and traditional relay control system?

(2) Is the implementation of the PLC program unique for a certain control task?

【Lab safety precautions】

(1) It is allowed to press the power button in order to prevent electric shock after the instructor confirms.

(2) Obey the wiring rules. The wires are connected in the order of ground first, and then the power cord. The order of disconnecting the wires is first the power cord, and then the ground.

(3) It is only allowed to disconnect the wires after turning off the power.

Chapter 4

Electronics Lab Parts

4.1 Common emitter amplifier circuits

【Lab objective】

(1) Understand the voltage gain, current gain, input impedance, output impedance and quiescent point.

(2) Understand the input characteristic curve and output characteristic curve.

(3) Learn the mlethods to test common emitter amplifier circuits.

【Lab devices】

Capacitors, resistors, NPN bipolar junction transistor, AC millivolt meter, function generator, oscilloscope, DC voltage power supply, and wires.

【Lab principle】

In a common emitter amplifier circuit as shown in Figure 4-1-1, the input is applied to base-emitter junction and the output is taken from emitter-collector terminal, emitter terminal is common for both input and output. R_{b1} and R_{b2} are known as biasing resistors.

Capacitors C_1 and C_2 are used as coupling capacitors to separate the AC signals from the DC biasing voltage. This ensures that the bias condition set up for the circuit to operate correctly is not affected by any additional amplifier stages, because the capacitors will only pass AC signals and block any DC components. Therefore, the output AC signal is then superimposed on the biasing of the following stages. In addition, a bypass capacitor C_3 is included in the emitter leg circuit. It is very helpful because any noise signal that may be presented in AC signal will be passed out from bypass capacitor. In other words, these capacitors are effectively open circuit components for DC biasing conditions, which means that the biasing currents and voltages are not affected by the addition of the capacitor maintaining a good Q-point stability.

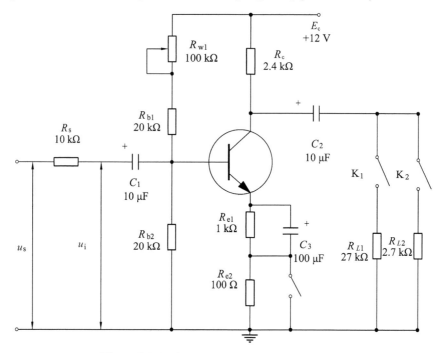

Figure 4-1-1　Common emitter amplifier circuit

As shown in Figure 4-1-2, the input impedance of an amplifier is the input impedance "seen" by the source driving the input of the amplifier. The input impedance can be expressed by

$$R_{in} = \frac{U_{in}}{U_s - U_{in}} R_s \qquad (4\text{-}1\text{-}1)$$

where U_s is the signal source, R_s is a known resistance, and U_{in} is the voltage across the amplifier.

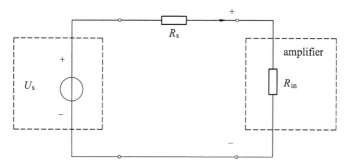

Figure 4-1-2　Input impedance of an amplifier

As shown in Figure 4-1-3, the amplifier becomes the source feeding the load when a load is connected to an amplifier. The output impedance can be expressed by

$$R_{out} = \frac{U_{out} - U_L}{U_L} R_L \qquad (4\text{-}1\text{-}2)$$

where U_{out} is the voltage across the amplifier, R_L is the load resistance, and U_L is the voltage across the load.

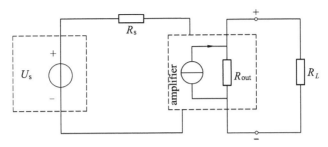

Figure 4-1-3 Output impedance of an amplifier

An ideal amplifier has an infinite input impedance and a zero value for the output impedance. Generally, an input impedance is high and an output impedance is low.

【Lab procedure】

(1) *Quiescent point*: Connect the circuit as shown in Figure 4-1-1, and switch on the DC voltage power supply after checking the circuit. Notice the input terminal set to zero. Measure the voltage U_{CE}, U_{BE}, U_{RE} and U_{RC} with the AC millivolt meter. Calculate the current I_C and I_B. Make records of the measured values, and fill them in the Table 4-1-1.

Table 4-1-1 Quiescent point

items	U_{CE}/V	U_{BE}/V	U_{RE}/V	U_{RC}/V	I_C/mA	I_B/mA
measured value						

(2) *Voltage gain*: The sinusoidal signal is generated by a function generator with a frequency and amplitude of 1 kHz and 10 mV, respectively. The input terminal is applied to the sinusoidal signal. Under three different load conditions (no load, R_{L1} and R_{L2}), observe the input and output signal waveform with the oscilloscope and draw them on the graph sheet. Make records of the measured values, and fill them in the Table 4-1-2.

Table 4-1-2 Voltage gain

load	U_i/mV	U_o/V	voltage gain
no load			
R_{L1}			
R_{L2}			

(3) *Input impedance*: Connect the circuit as shown in Figure 4-1-2, switch on the function generator after checking the circuit. Observe the input and output signal waveforms with the oscilloscope and draw them on the graph sheet. Make records of the measured values, and fill

them in the Table 4-1-3.

(4) *Output impedance*: Connect the circuit as shown in Figure 4-1-3, switch on the function generator after checking the circuit. Observe the input and output signal waveforms with the oscilloscope and draw them on the graph sheet. Make records of measured values, and fill them in the Table 4-1-3.

Table 4-1-3 Input impedance and output impedance

items	R_{in}	R_{out}
R_s		-
R_L	-	

【Pre-lab questions】

(1) Preview the concepts of voltage gain, current gain, and quiescent point.
(2) Preview the concepts of input impedance and output impedance.
(3) Preview the features of common emitter amplifier circuits.

【Post-lab questions】

(1) What causes the saturation distortion in common emitter amplifier circuits?
(2) What causes the cut off distortion in common emitter amplifier circuits?
(3) Which parameters affect the quiescent point in common emitter amplifier circuits?
(4) What is the voltage gain in common emitter amplifier circuits?

4.2 Long-tail pairs differential amplifier circuits

【Lab objective】

(1) Understand the principle of long-tail pairs differential amplifier circuits.

(2) Understand the differential gain, common mode gain, and common mode rejection ratio.

(3) Learn the methods to test the long-tail pairs differential amplifier circuits.

【Lab devices】

Resistors, NPN bipolar junction transistors, function generator, AC millivolt meter, oscilloscope, a DC voltage source, and wires.

【Lab principle】

In the long-tail pairs differential amplifier circuit as shown in Figure 4-2-1, the two separate transistors (T_1 and T_2) possess similar parameter: common emitter resistor R_e, common positive supply U_{cc}, and common negative supply U_{ee}. There are two inputs (u_{i1} and u_{i2}) and two outputs (u_{o1} and u_{o2}). Therefore, four operating modes are briefly given as follows.

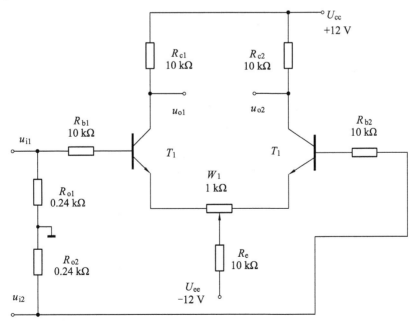

Figure 4-2-1 The long-tail pairs differential amplifier circuit

(1) *Dual input balanced output*: two inputs are given from two separate transistors and two

outputs are taken from two separate transistors.

(2) *Dual input unbalanced output*: two inputs are given from two separate transistors and an output is taken from only a single transistor.

(3) *Single input balanced output*: single input is provided from only a single transistor and the outputs are taken from two separate transistors.

(4) *Single input unbalance output*: single input is provided from only a single transistor and an output is taken from only a single transistor.

The difference is defined as

$$u_d(t) = \frac{u_2(t) - u_1(t)}{2} \tag{4-2-1}$$

The common-mode signal is defined as

$$u_{cm}(t) = \frac{u_2(t) + u_1(t)}{2} \tag{4-2-2}$$

In the differential amplifier circuit, the output is expressed by

$$u_{out}(t) = A_d u_d(t) + A_{cm} u_{cm}(t) \tag{4-2-3}$$

where A_d is the differential gain and A_{cm} is common-mode gain. An ideal differential amplifier has zero common-mode gain. In other words, the output of an ideal differential amplifier is independent of the common-mode (i.e., average) of the two input signals. The common-mode rejection ratio (CMRR) is defined as

$$CMRR = 10\lg\frac{|A_d|^2}{|A_{cm}|^2} \tag{4-2-4}$$

The common-mode rejection ratio (CMRR) is used to indicate the quality of a differential amplifier. In addition, the long-tail pairs differential amplifier circuits have low input impedance and high output impedance.

【Lab procedure】

Firstly, zeroing the differential amplifier circuit. When the two input terminals (u_{i1} and u_{i2}) are grounded, and the two output signals (u_{o1} and u_{o2}) are adjusted to zero by rotating the potentiometer R_w.

(1) *Quiescent point*: Connect the circuit as shown in Figure 4-2-1 and switch on the DC voltage power supply after checking the circuit. Measure the voltages of transistors (T_1 and T_2) with AC millivolt meter. Make records of the measured values, and fill them in Table 4-2-1.

Table 4-2-1 Quiescent point

items	U_{C1}/mV	U_{E1}/mV	U_{B1}/mV	U_{C2}/mV	U_{E2}/mV	U_{B2}/mV
measured value						

(2) *Differential gain*: The sinusoidal signal is generated by a function generator with a frequency and amplitude of 80 Hz and 0.3 V, respectively. The input terminal of differential amplifier circuit is applied to the sinusoidal signal. Measure the output voltages, make records of

the measured values, and fill them in Table 4-2-2.

Table 4-2-2 Differential gain

items	U_{o1}/mV	U_{o2}/mV	A_{d1}	A_{d2}	A_d
measured value					

(3) *Common-mode gain*: The sinusoidal signal is generated by a function generator with a frequency and amplitude of 80 Hz and 0.3 V, respectively. The input terminal of long-tail pairs differential amplifier circuit is applied to the sinusoidal signal. Measure the output voltages of long-tail pairs differential amplifier circuits. Make records of the measured values, and fill them in Table 4-2-3.

Table 4-2-3 Common-mode gain

items	U_{o1}/mV	U_{o2}/mV	A_{cm1}	A_{cm2}	A_{cm}
measured value					

【Pre-lab questions】

(1) Preview the concepts of differential gain and common mode gain.

(2) Preview the concepts of common mode rejection ratio.

(3) Preview the principle of differential amplifier circuits.

【Post-lab questions】

(1) What is the role of the resistor R_e in long-tail pairs differential amplifier circuits?

(2) Why are the symmetry parameters used in long-tail pairs differential amplifier circuits?

(3) What are the relationships between the input signals and the output signals of the long-tail pairs differential amplifier circuits?

(4) What's the reason of the zero-drift of amplifiers?

4.3 Negative feedback amplifier circuits

【Lab objective】

(1) Understand the structure of negative feedback amplifier circuits.
(2) Understand the performance of a negative feedback amplifier circuits.
(3) Learn the methods to test negative feedback amplifiers.

【Lab devices】

Resistors, capacitors, transistors, function generator, oscilloscope, a DC voltage source, millivolt meters, and wires.

【Lab principle】

Feedback amplifiers can be used to control the circuit gain, response, bandwidth, and signal distortion. As shown in Figure 4-3-1, a feedback amplifier is an amplifier in which a portion of the output is returned to the input. Depending on the polarity relationship between the source and the feedback signal, there are positive and negative feedback amplifiers. In a positive feedback amplifier, the input signal will be increased because the source and feedback signals are in the same phase. However, the input signal will be decreased because the source and feedback signals are out of phase.

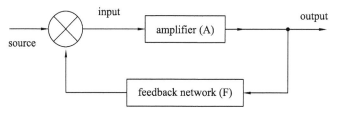

Figure 4-3-1 The block diagram of a feedback amplifier

In Figure 4-3-1, the input and output voltage of the amplifier note U_i and U_o, respectively. The gain of the amplifier is represented as A. The negative feedback network extracts a voltage $U_f = -\beta U_o$. The input voltage is expressed by

$$U_i = U_s + U_f = U_s - \beta U_o \qquad (4\text{-}3\text{-}1)$$

$$A = \frac{U_o}{U_i} \qquad (4\text{-}3\text{-}2)$$

where β is the amplifier gain. The ratio of output voltage to the applied signal voltage is the gain of the feedback amplifier, which is expressed by

$$A_f = \frac{U_o}{U_s} = \frac{A}{1 + \beta A} \qquad (4\text{-}3\text{-}3)$$

The feedback signal is usually voltage and current. According to the connection of the feedback signal, there are voltage-series feedback, voltage-shunt feedback, current-series feedback, and current-shunt feedback. The effects of feedback amplifiers are listed in Table 4-3-1.

Table 4-3-1　The effects of feedback amplifiers

parameter	voltage-series	voltage-shunt	current-series	current-shunt
voltage gain	small	small	small	small
bandwidth	large	large	large	large
input resistance	large	small	large	small
output resistance	small	small	large	large
harmonic distortion	small	small	small	small
noise	small	small	small	small

【Lab procedure】

(1) Connect the circuit as shown in Figure 4-3-2, and switch on the power after checking the circuit.

(2) The input signal of transistor T_1 is generated by the output of the function generator, which is set to sinusoidal signal. The frequency of function signal is 1 kHz and the amplitude is 5 mV. Observe the output signal waveforms of transistor T_1 and T_2 with the oscilloscope.

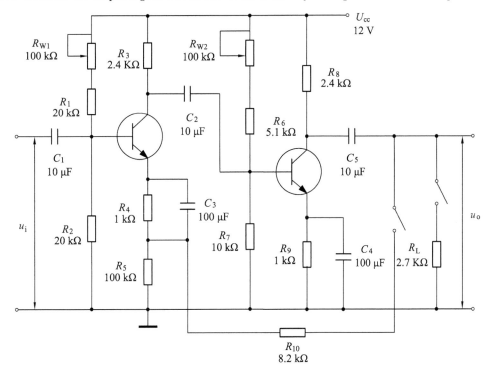

Figure 4-3-2　The negative feedback amplifier circuit

(3) Measure the gain of transistor T_1 with and without transistor T_2. Observe the results,

make records of the measured values, and fill them in Table 4-3-2.

Table 4-3-2 The gain of transistor T_1

items	with transistor T_2	without transistor T_2
U_{o1}		
A_1		

(4) Measure the gain of transistor T_2 with and without load. Observe the results, make records of the measured values, and fill them in Table 4-3-3.

Table 4-3-3 The gain of transistor T_2

items	with load	without load
U_{o2}		
A_2		

【Pre-lab questions】

(1) Preview the concept of feedback in the circuits.
(2) Preview the principle of the negative feedback amplifier circuits.
(3) Preview the features of the negative feedback amplifier circuits.

【Post-lab questions】

(1) What are the roles of the negative feedback amplifier circuits?
(2) What is the difference between feedback circuit with and without load?
(3) What are the characteristics of the secondary amplifier circuit?
(4) How many kinds of feedback circuits?

4.4 Integrated operational amplifier circuits

【Lab objective】

(1) Understand the structure and performance of integrated operational amplifiers.

(2) Understand the function of the summing, differential, differentiator and integrator operational circuits.

(3) Learn to use the integrated operational amplifier, such as LM741.

【Lab devices】

Resistors, integrated operational amplifier LM741, function generator, oscilloscope, a DC voltage power supply, AC millivolt meter, and wires.

【Lab principle】

An operational amplifier is commonly abbreviated by op amp, basically an amplifying device with external feedback components such as resistors and capacitors between its output and input terminals. The typical op amp is of a 3-terminal device with 2-inputs (an inverting input is marked with a " $-$ " sign and a non-inverting input is marked with a " $+$ " sign) and 1-output, which need a DC voltage power supply. In addition, it has an infinite input impedance and zero output impedance. The application fields of op amps include signal condition, filters, and mathematical operations.

As shown in Figure 4-4-1, the output voltage of inverting operational circuit is expressed by

$$u_o = -\frac{R_F}{R_1} u_i \qquad (4\text{-}4\text{-}1)$$

where u_i is the input signal, u_o is the output signal, R_1 is the input resistor, and R_F is the feedback resistor. The negative '$-$' sign indicates that there is a phase shift of π between input and output because of the inverting input terminal of the op-amp. As well as constructing the inverting amplifier, there are the non-inverting amplifier that produces an output signal which is "in-phase" with the input.

As shown in Figure 4-4-2, the input signal is applied to the inverting terminal, called inverting summing amplifiers. The output voltage of summing operational circuit is expressed by

$$u_o = -\left(\frac{u_{i1}}{R_{11}} + \frac{u_{i2}}{R_{12}} + \frac{u_{i3}}{R_{13}}\right) R_F \qquad (4\text{-}4\text{-}2)$$

where u_{i1}, u_{i2} and u_{i3} are the input signals; R_{11}, R_{12} and R_{13} are the input resistors. As well as constructing inverting summing amplifiers, use the non-inverting input of the operational amplifier to produce a non-inverting summing amplifier.

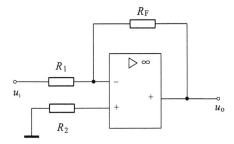
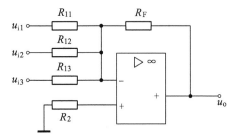

Figure 4-4-1 the inverting operational circuit   **Figure 4-4-2 the summing operational circuit**

A differential operation is the difference between two input signals. A differential operational amplifier has an inverting input and a non-inverting input. As shown in Figure 4-4-3, if $R_2 = R_1$ and $R_3 = R_F$, then, the output voltage of the differential amplifier circuit is expressed by

$$u_o = \frac{R_F}{R_1}(u_{i2} - u_{i1}) \tag{4-4-3}$$

The differential amplifier is a combination of both inverting and non-inverting amplifiers. It uses a negative feedback connection to control the differential voltage gain.

As shown in Figure 4-4-4, the output voltage of integrator operational circuit is expressed by

$$u_o = -\frac{1}{R_1 C}\int_0^t u_i \mathrm{d}t \tag{4-4-4}$$

where the product term $R_1 C$ is known as the time constant of the integrator. An integrating circuit is usually used to generate a ramp wave from a square wave input.

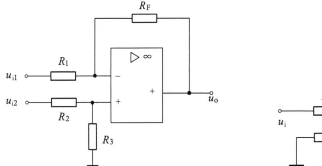
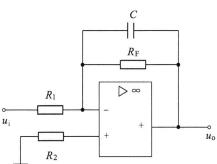

Figure 4-4-3 The differential operational circuit   **Figure 4-4-4 The integrator operational circuit**

As shown in Figure 4-4-5, the output voltage of differentiator operational circuit is expressed by

$$u_o = -R_F C \frac{\mathrm{d}u_i}{\mathrm{d}t} \tag{4-4-5}$$

where the product term $R_F C$ is known as the time constant of the differentiator. Differentiating circuits are most commonly designed to operate on triangular and rectangular signals.

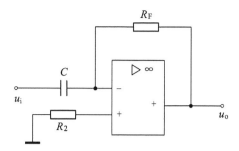

Figure 4-4-5 The differentiator operational circuit

【Lab procedure】

As shown in Figure 4-4-6, the integrated operational amplifier LM741 is used in this lab. Firstly, carefully identify the 8 pins of the chip LM741 which are listed in Table 4-4-1.

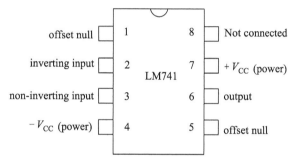

Figure 4-4-6 The diagram of LM741 pins

Table 4-4-1 The LM741 pins

No.	name	details
1	offset null	pin 1 is for external input offset voltage adjustment
2	inverting input	pin 2 is applied to inverting input
3	non-inverting input	pin 3 is applied to non-inverting input voltage
4	$-V_{cc}$	pin 4 is applied to negative DC power supply
5	offset null	Pin 5 is for external input offset voltage adjustment
6	output	pin 6 is applied to output voltage
7	$+V_{cc}$	pin 7 is applied to positive DC power supply
8	Not connected	pin 8 is for no connection and it is to be kept open

(1) *Zeroing IC*: Before the lab, the LM741 needs to be zeroed in order to reduce the zero-drift. The zeroing circuit is illustrated in Figure 4-4-7. When the input terminal is grounded, and the output signal is adjusted to zero by rotating the potentiometer R_W.

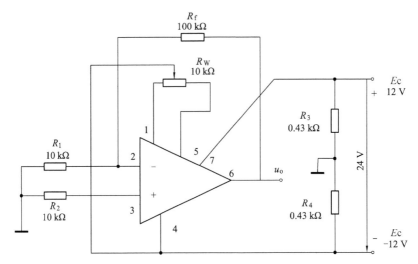

Figure 4-4-7   The zeroing circuit

(2) *Inverting*: Connect the circuit according to Figure 4-4-1. The pin 2 is applied to a DC voltage with the amplitude of 0.5 V. The output voltage of pin 6 is measured with an AC millivolt meter. Make records of the measured values, and fill them in Table 4-4-2.

Table 4-4-2   The inverting, differential and sum operations

items	u_i	u_o	A
inverting	$u_i = 0.5$ V		
differential	$u_{i1} = 1.0$ V; $u_{i2} = 0.5$ V;		—
sum	$u_{i1} = 1.0$ V; $u_{i2} = 0.5$ V; $u_{i3} = 0.3$ V		—

(3) *Summing*: Connect the circuit according to Figure 4-4-2, and switch on the power supply after checking the circuit. Test the sum operations. Make records of the measured value, and fill it in Table 4-4-2.

(4) *Differential*: Connect the circuit according to Figure 4-4-3, and switch on the power supply after checking the circuit. Test the differential operations. Make records of the measured value, and fill it in Table 4-4-2.

(5) *Integrator*: Connect the circuit according to Figure 4-4-4, and switch on the power supply after checking the circuit. The pin 2 is applied to a rectangle signal with a frequency and amplitude of 1 kHz and 0.5 V, respectively. Observe the output waveform with an oscilloscope and draw the waveform on the graph sheet.

(6) differentiator: Connect the circuit according to Figure 4-4-5, and switch on the power supply after checking the circuit. The pin 2 is applied to a rectangle signal with a frequency and amplitude of 1 kHz and 0.5 V, respectively. Observe the output waveform with an oscilloscope and draw the waveform on the graph sheet.

【Pre-lab questions】

(1) Preview the principle of the inverting, summing, and differential circuits.

(2) Preview the principle of the differentiator and integrator operation circuits.

(3) Preview the structure and performance of integrated operational amplifier.

【Post-lab questions】

(1) Why should the amplifier be connected in a closed loop while zeroing?

(2) What is the role of the resistor R_F in the integrator amplifier?

(3) What is the role of the resistor R_1 in the differential amplifier?

4.5 Applications of integrated operational amplifiers

【Lab objective】

(1) Understand the principle of a hysteresis voltage comparator.
(2) Understand the principle of a square wave generator.
(3) Learn the methods to test an integrated operational circuit.

【Lab devices】

Resistors, integrated operational amplifier LM741, function generator, oscilloscope, a DC voltage supply, AC millivolt meter, and wires.

【Lab principle】

An oscillator is a device used to generate repetitive signals, which produces an output signal without an input signal. There are harmonic type oscillators and relaxation type oscillators. The harmonic oscillator mainly produces a sine wave output. However, the relaxation oscillator produces non-sine wave outputs. As shown in Figure 4-5-1, a sine wave oscillation consists of an op amp, $R_3 C_1$ in parallel circuit and $R_4 C_2$ series circuit. When R_3 is equal to R_4, and C_1 is equal to C_2, the frequency of the sine wave is expressed by

$$f_0 = \frac{1}{R_3 C_1} = \frac{1}{R_4 C_2} \tag{4-5-1}$$

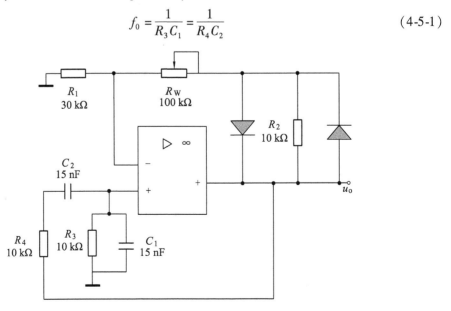

Figure 4-5-1　A sine wave oscillation

As shown in Figure 4-5-2, a hysteresis comparator is operated by applying a positive feedback to the comparator. In this circuit, the resistor R_1 is connected to the inverting terminal of the operational amplifier, the resistors R_2 and R_3 are connected to the non-inverting terminal of the operational amplifier, and the resistor R_4 is a current limiting resistor that is used to reduce the current in a circuit.

The potential difference between the high and low output voltages and the feedback resistor are adjusted to change the voltage. The width of variation in the reference voltage is the hysteresis width. Output voltage amplitude can be adjusted by bidirectional voltage regulator. The threshold voltage is expressed by

$$\pm U_T = \pm \frac{R_2}{R_2 + R_3} U_z \qquad (4\text{-}5\text{-}2)$$

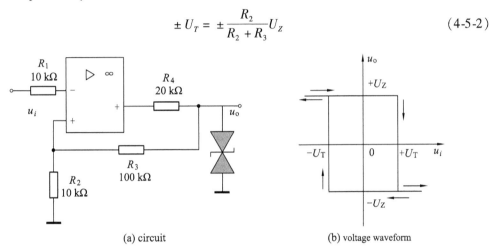

Figure 4-5-2 A hysteresis comparator

As shown in Figure 4-5-3, square wave generator consists of a hysteresis comparator and RC series circuits.

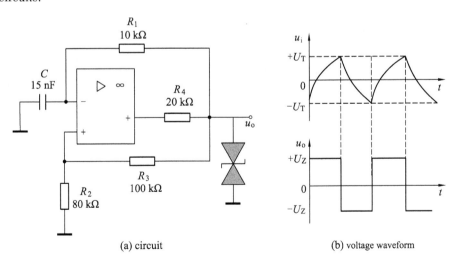

Figure 4-5-3 A square wave generator

The capacitor C and resistor R_1 are connected to the inverting terminal of the operational amplifier, the resistors R_2 and R_3 are connected to the non-inverting terminal of the operational amplifier, and the resistor R_4 is a current limiting resistor that is used to reduce the current in a circuit. By charging and discharging the capacitor, the square wave is automatically turned over. The period of the square wave is expressed by

$$T = 2CR_1 \ln\left(1 + 2\frac{R_3}{R_2}\right) \qquad (4\text{-}5\text{-}3)$$

【Lab procedure】

(1) *A sine wave oscillation*: Connect the circuit as shown in Figure 4-5-1, and switch on the power supply after checking the circuit. Observe the output waveform using an oscilloscope, and draw them on the graph sheet.

(2) *A hysteresis comparator*: Connect the circuit as shown in Figure 4-5-2, and switch on the power supply after checking the circuit. Observe the output waveform using an oscilloscope, and draw them on the graph sheet.

(3) *A square wave generator*: Connect the circuit as shown in Figure 4-5-3, and switch on the power supply after checking the circuit. Observe the output waveform using an oscilloscope, and draw them on the graph sheet.

【Pre-lab questions】

(1) Preview the principle of a hysteresis voltage comparator.
(2) Preview the principle of a square wave generator.
(3) Preview the features of bidirectional voltage regulators.

【Post-lab questions】

(1) What are the application fields of the integrated op amps?
(2) What is the role of bidirectional voltage regulator in the hysteresis voltage comparator?
(3) What is the role of negative feedbackin the square wave generator?
(4) What is the principle of an oscillator?

4.6 Rectified DC power supplies

【Lab objective】

(1) Understand the principle of bridge rectifiers and regulated voltage circuits.

(2) Understand the structure of a rectified DC power supply.

(3) Learn the methods to test a rectified DC power supply.

【Lab devices】

Capacitors, integrated voltage regulator 7812, integrated voltage regulator 7912, oscilloscope, a DC voltmeter, and wires.

【Lab principle】

In the transistor amplifier circuit, the DC bias voltage is provided by the DC power supply. As we well known, digital electronics are composed of semiconductors, such as transistors and diodes. Thus, the DC power supply is a vital device in digital electronics world. As shown in Figure 4-6-1, a rectified DC power supply consists of step-down transformer, bridge rectifier, filter, and regulated voltage. The main functions are briefly given as follows.

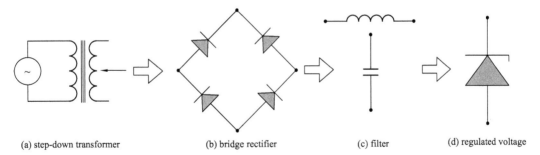

(a) step-down transformer　　(b) bridge rectifier　　(c) filter　　(d) regulated voltage

Figure 4-6-1　The diagram of rectified DC power supply

(1) Using a step-down transformer, the 220 V AC voltage from a wall outlet is converted to low AC voltage. The waveform from the output of a step-down is illustrated in Figure 4-6-2a.

(2) A diode-bridge rectifier consists of four or more diodes depending on the type of bridge rectifier, also called full-wave rectifier. Since the diode is a unidirectional device that allows the current to flow in one direction only, the low AC voltage from the step-down transform is converted to DC voltage. As shown in Figure 4-6-2b, the DC output from a full-wave rectifier is varying and unstable.

(3) Because the output of the diode bridge rectifier has pulsating nature, it must be filtered to produce a pure DC signal. The smoothing capacitor acts to smooth out fluctuations in a signal,

so that there is less ripple. As shown in Figure 4-6-2c, the output from the filter is still a slight ripple.

(4) A voltage regulator serves to further smooth out the fluctuating signal and maintains the output voltage to a constant level, so that the output from a voltage regulator is a perfectly smoothing DC signal, which is illustrated in Figure 4-6-2d.

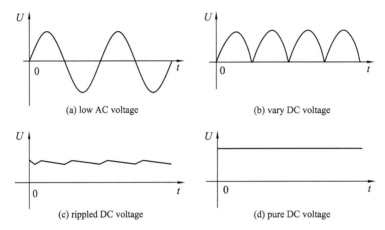

Figure 4-6-2 The waveform of a rectified DC power supply

【Lab procedure】

Connect the circuit as shown in Figure 4-6-3, and switch on the AC voltage power supply after checking the polarity of the step-down transformer and the integrated voltage regulator.

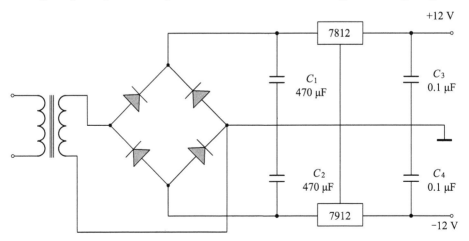

Figure 4-6-3 The rectified DC power supply

(1) Connect the circuit between the step-down transformer and the diode-bridge rectifier without the capacitors filter. Observe the output waveform from the diode-bridge rectifier with an oscilloscope and draw them on the graph sheet. Make records of the measured values and fill them in the Table 4-6-1.

Table 4-6-1 Diode-bridge rectifier

items	$U_{1\text{primary}}$/V	$U_{2\text{ second}}$/V	U_{bridge}
measured value			

(2) Connect the circuit between the step-down transformer and the capacitor filters without the voltage regulator. Observe the output waveforms from the capacitor filters with an oscilloscope and draw them on the graph sheet. Make records of the measured values, and fill them in Table 4-6-2.

Table 4-6-2 Capacitors filter

items	U_{c1}/V	U_{c2}/V
measured value		

(3) Connect the total circuit based on three different loads ($R_L = \infty /3.6 \text{ k}\Omega/1.2 \text{ k}\Omega$), measure the output of the voltage regulator. Observe the output waveforms from the voltage regulator with an oscilloscope and draw them on the graph sheet. Make records of the measured values, and fill them in Table 4-6-3.

Table 4-6-3 Voltage regulator

items	∞	3.6 kΩ	1.2 kΩ
measured value			

【Pre-lab questions】

(1) Preview the principle of the rectified DC power supply.
(2) Preview the principle of the diode-bridge rectifiers.
(3) Preview the structure and performance of the rectified DC power supply.

【Post-lab questions】

(1) What are the advantages and disadvantages of bridge rectifier?
(2) What is the role of the capacitors in the rectified DC power supply?
(3) What is the role of the diodes in the rectified DC power supply?

4.7 Test logic function of basic gates circuits

【Lab objective】

(1) Understand the principle of basic logic gates, such as AND, OR, NOT, NAND, NOR, XOR.
(2) Understand the principle and function of a tri-state gate.
(3) Learn to design AND and XOR circuits with NAND gates.
(4) Learn the methods to test integrated logic circuit chips.

【Lab devices】

Signal generator, LED, oscilloscope, regulated power supply, wires, integrated chips 74LS00 and 74LS125.

【Lab principle】

A logic gate is a building block of a digital circuit. Most logic gates have two inputs and one output, they are based on Boolean algebra. There are seven basic logic gates: AND, OR, XOR, NOT, NAND, NOR, and XNOR. However, logic gates like NAND, NOR are used in daily applications for performing logic operations. The gates are manufactured using semiconductor devices like BJTs, Diodes or FETs.

Different gates are constructed using integrated circuits. Digital logic circuits are manufactured depending on the specific circuit technology or logic families such as TTL (transistor transistor logic) and CMOS (complementary metal oxide semiconductor logic). These integrated circuits are small silicon semiconductors sheets called chips, containing the electronic components for the logic gates. The chip is mounted in a plastic container, and connections are welded to external pins which may range from 14 in a small IC package to 64 or more in a large one.

Basic TTL NAND gate circuit is illustrated in Figure 4-7-1. Transistor Q_1 has one collector, one base and two emitters. The two diodes named as D_1 and D_2 are used to limit the input voltages. Acceptable TTL gate input and output signal level is illustrated in Figure 4-7-2.

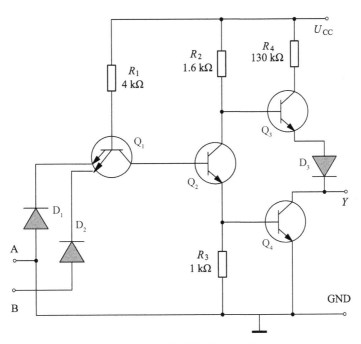

Figure 4-7-1　TTL NAND gate circuit

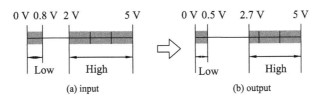

Figure 4-7-2　Acceptable TTL gate signal level

As shown in Figure 4-7-3, a tri-state gate is a digital device, which has a data input and a control input. It is capable of three different outputs: high, low and disconnected. Therefore, the control input acts like a valve. When the control input is active, the output is the input. When the control input is not active no electrical current flows through, so the tri-state gate is in high impedance state. Thus, even if A is 0 or 1, that value does not flow through the gate.

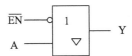

Figure 4-7-3　A tri-state gate symbol

【Lab procedure】

As shown in Figure 4-7-4, carefully identify the 14 pins of the chip 74LS00. "74" means 7400 TTL series, "LS" means low schottky type. There are 4 NAND gates in the chip 74LS00. The two inputs and the one output of NAND gate 1 are 1A, 1B and 1Y (pins 1, 2, and 3), respectively. Therefore, the chip 74LS00 is also called quad 2 input NAND gate.

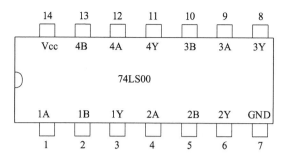

Figure 4-7-4　The diagram of the chip 74LS00 pins

As shown in Figure 4-7-5, there are 4 tri-state buffers in the chip 74LS125. The two inputs and the one output of tri-state gate 1 are $1\overline{E}, 1A$, and $1Y$ (pins 1, 2, and 3), respectively. Therefore, the chip 74LS125 is also called quad tri-state buffer.

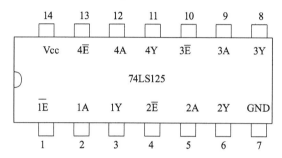

Figure 4-7-5　The diagram of the chip 74LS125 pins

In the lab, signal generator creates the TTL normal signal with a frequency of 1 Hz and we test the circuit output with LED light.

(1) *AND logic operation*: As shown in Figure 4-7-6, the circuit of AND operation is implemented by NAND gates. Connect the circuit using the integrated chip 74LS00 and verify the function of AND logic operation. Make records of the measured values, and fill them in Table 4-7-1.

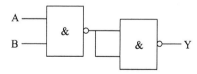

Figure 4-7-6　AND operation with NAND gates

(2) *XOR logic operation*: As shown in Figure 4-7-7, the circuit of XOR operation is implemented by NAND gates. Connect the circuit using the integrated chip 74LS00 and verify the function of XOR logic operation. Make records of the measured values, and fill them in Table 4-7-1.

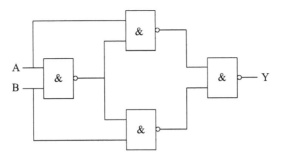

Figure 4-7-7 XOR operation with NAND gates

Table 4-7-1 The logic state table

AND			XOR			Tri-state gate		
A	B	Y	A	B	Y	A	B	Y
0	0		0	0		0	0	
0	1		0	1		0	1	
1	0		1	0		1	0	
1	1		1	1		1	1	

(3) *A tri-state gate*: the circuits of testing tri-state gate is illustrated in Figure 4-7-8. Connect the circuit using the integrated chip 74LS125 and verify the circuit according to the results. Make records of the measured values, and fill them in Table 4-7-1.

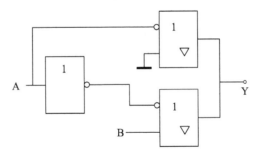

Figure 4-7-8 Test the tri-state gate

【Pre-lab questions】

(1) Preview the principle of AND, OR, and NOT logic gates.

(2) Preview the principle of NAND, NOR, and XOR logic gates.

(3) Preview the pins and data sheets of the integrated chip 74LS00 and 74LS125.

【Post-lab questions】

(1) How to verify the functions of logic gates OR, NOT, and NOR?

(2) What is the role of the tri-state gate in circuits?

(3) What is the acceptable level ranges of TTL gate input and output signals?

4.8 Test combinational logic circuits

【Lab objective】

(1) Understand the features of the combinational logic circuits.
(2) Learn to design the half adder and full adder.
(3) Learn to design the data selector.

【Lab devices】

Signal generator, LED, oscilloscope, regulated DC power supply, wires, integrated chip 74LS10 and 74LS86.

【Lab principle】

In contrast to a sequential logic circuit, a combinational logic circuit can be defined as a digital logic circuit using a Boolean circuit, where the output of the logic circuit is only a pure function of the present input. In a combinational logic circuit, there is no memory or feedback loop. The typical applications are half adder, full adder, and data selector.

An adder is a digital logic circuit in electronics that implements addition of numbers. Adders are classified into two types: half adder and full adder. The half adder circuit has two inputs: A and B, which add two input digits and generate a carry and a sum. As shown in Figure 4-8-1, there are 5 NAND gates.

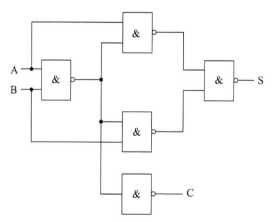

Figure 4-8-1 A half adder with NAND gates

The main difference between a half-adder and a full-adder is that the full-adder has three inputs and two outputs. The first two inputs are A and B and the third input is an input carry. The outputs are a carry and a sum, respectively. As shown in Figure 4-8-2, there are 3 NAND gates and 2 exclusive-OR gates.

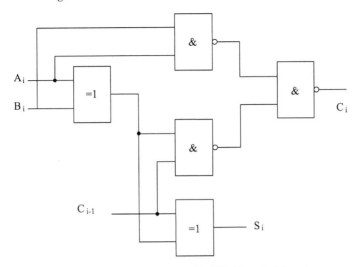

Figure 4-8-2　A full adder with NAND and OR gates

In digital electronics, selecting multiple data sources can be performed by combinational logic circuits. The names "data selector" and "multiplexer" are commonly interchanged, with multiplexers called data selectors and vice versa. The multiplexer is a combinational logic circuit designed to switch one of several input lines through to a single common output line by the application of a control signal. As shown in Figure 4-8-3, there are 2 NAND gates with 1 input, 4 NAND gates with 3 inputs and 1 NAND gate with 4 inputs.

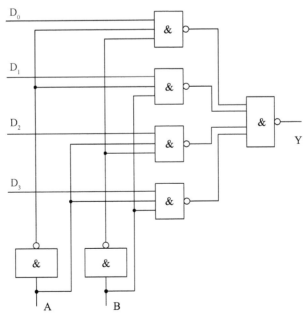

Figure 4-8-3　The data selector with NAND gates

【Lab procedure】

As shown in Figure 4-8-4, there are triple 3-input NAND gates in the chip 74LS10. The three inputs and the one output of NAND gate 1 are $1A$, $1B$, $1C$, and $1Y$ (pins 1, 2, 13, and 12), respectively.

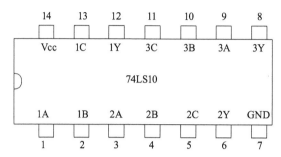

Figure 4-8-4　The diagram of the chip 74LS10 14 pins

As shown in Figure 4-8-5, there are quad exclusive-OR gates in the chip 74LS86. The two inputs and the one output of OR gate 1 are $1A$, $1B$, and $1Y$ (pins 1, 2, and 3), respectively.

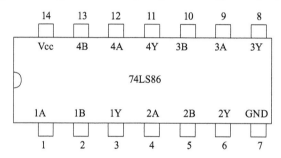

Figure 4-8-5　The diagram of the chip 74LS86 14 pins

(1) *A half adder*: As shown in Figure 4-8-1, the circuit of a half adder is implemented by NAND gates. Connect the circuit using the integrated chip 74LS00 and verify the half adder circuit according to the test results. Make records of the measured values, and fill them in Table 4-8-1.

(2) *A full adder*: As shown in Figure 4-8-2, the circuit of a full adder is implemented by NAND and exclusive-OR gates. Connect the circuit using the integrated chips 74LS10 and 74LS86, and verify the full adder circuit according to the test results. Make records of the measured values, and fill them in Table 4-8-1.

(3) *A data selector*: As shown in Figure 4-8-3, the circuit of a half adder is implemented by NAND gates. Connect the circuit using the integrated chip 74LS86 and verify the data selector circuit according to the test results. Make records of the measured values, and fill them in Table 4-8-1.

Table 4-8-1 The logic state table

half adder			full adder					data selector			
A	B	Y	A_i	B_i	C_{i-1}	S_i	C_i	A	B	$D_0D_1D_2D_3$	Y
0	0		0	0	0			0	0	0 × × ×	
0	1		0	0	1			0	0	1 × × ×	
1	0		0	1	0			0	1	×0 × ×	
1	1		0	1	1			0	1	×1 × ×	
×	×		1	0	0			1	0	× ×0 ×	
×	×		1	0	1			1	0	× ×1 ×	
×	×		1	1	0			1	1	× × ×0	
×	×		1	1	1			1	1	× × ×1	

【Pre-lab questions】

(1) Preview the principle of the half adder and full adder.

(2) Preview the principle of the data selector.

(3) Preview the pins of the integrated chips 74LS10 and 74LS86.

【Post-lab questions】

(1) How to verify the relationship between logic input and logic output in the lab?

(2) What is the role of the data selector in circuits?

(3) What are the application fields of the data selector?

(4) What is the difference between a half adder and a full adder?

4.9 Design combinational logic circuits

【Lab objective】

(1) Understand the principle and function of basic combinational logic circuits.

(2) Learn to analyze specific problems using the basic combinational logic circuits.

(3) Learn to design a blood transfusion test by the combinational logic circuits.

【Lab devices】

Signal generator, LED, oscilloscope, regulated power supply, wires, integrated chips 74LS00, and 74LS151.

【Lab principle】

Combinational circuits are electronic digital circuits with N inputs and M outputs using logic gates, which have no memory function and their present state is not affected in any way by the previous state. The state of the combinational circuit is completely dependent on the circuit inputs at the time: logic states 0 and 1. The design procedure for combinational logic circuits starts with the problem specification and comprises the following steps:

(1) Determine required number of inputs and outputs from the specifications.

(2) Derive the truth Table for each of the outputs based on their relationships to the input.

(3) Simplify the Boolean expression for each output. Use Karnaugh Maps or Boolean algebra.

(4) Draw a logic diagram that represents the simplified Boolean expression. Verify the design using the logic circuit.

【Lab procedure】

There are many blood groups in the human population including ABO, Rh, Kidd and Lewis. In an ABO blood group, there are A, B, AB and O types. Avoiding ABO incompatible transfusions, the rules of blood transfusions are listed in Table 4-9-1.

Table 4-9-1 Blood transfusions

blood type of donor	blood type of recipient			
	A	B	AB	O
A	✓	×	✓	×
B	×	✓	✓	×
AB	×	×	✓	×
O	✓	✓	✓	✓

Chapter 4   Electronics Lab Parts

The logic state Table of blood transfusions is listed in Table 4-9-2, where A type blood is "00", B type blood is "01", AB type blood is "10", and O type blood is "11". The chip 74LS151 is 8-input multiplexer. Design a combinational logic circuit to implement the human blood transfusions using chip 74LS151 and 74LS00. Draw the logic circuit and test it. Observe the results and make records.

Table 4-9-2   The logic state table

blood type of donor		blood type of recipient		output
0	0	0	0	1
0	0	0	1	0
0	0	1	0	1
0	0	1	1	0
0	1	0	0	0
0	1	0	1	1
0	1	1	0	1
0	1	1	1	0
1	0	0	0	0
1	0	0	1	0
1	0	1	0	1
1	0	1	1	0
1	1	0	0	1
1	1	0	1	1
1	1	1	0	1
1	1	1	1	1

【Pre-lab questions】

(1) Preview the design steps of the combinational logic circuit.

(2) Preview the principle of the basic gate circuit.

(3) Preview the principle of integrated chips.

【Post-lab questions】

(1) How to implement a blood transfusion test by the combinational logic circuit?

(2) What are the features of AND, OR, and NAND gates?

(3) What are the competition and risk in combinational logic circuits?

4.10 Test logic functions of flip flops

【Lab objective】

(1) Understand the principle and function of RS, D and JK type flip flops.
(2) Learn to test the methods of RS, D and JK flip flops.
(3) Learn to transform the methods among the RS, D and JK flip flop.

【Lab devices】

Signal generator, LED, oscilloscope, regulated DC power supply, wires, integrated chips 74LS112, 74LS74 and 74LS00.

【Lab principle】

A flip flop is a bi-stable device, which is divided into RS, JK and D flip flop types. As shown in Figure 4-10-1, the basic RS-type flip flop is a one-bit memory bi-stable device. It has two inputs, one is called "SET" which will set the device (output = 1) and is labeled "S". The other is known as "RESET" which will reset the device (output = 0) and is labeled as "R".

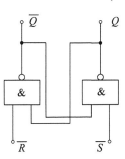

Figure 4-10-1 The logic symbol of RS type flip flop

The JK-type flip flop is considered to be a universal flip-flop circuit. The JK flip flop name has been kept on the inventor name of the circuit known as Jack Kilby who is a Texas instrument engineer. The characteristic equation of the JK type flip flop is expressed by

$$Q(t+1) = J\overline{Q}(t) + \overline{K}Q(t) \qquad (4\text{-}10\text{-}1)$$

The JK flip flop has four possible input combinations because of the addition of the clocked input, which is illustrated in Figure 4-10-2. The four inputs are "logic 1", "logic 0", "No change" and "Toggle". The truth table of negative-edge JK flip flop is listed in Table 4-10-1.

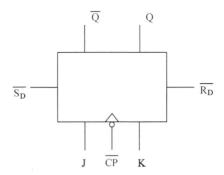

Figure 4-10-2 The logic symbol of JK type flip flop

Table 4-10-1 Truth table of negative-edge JK flip flop

	inputs				outputs	
set	reset	clock	J	K	Q	$\overline{Q}$
0	1	×	×	×	1	0
1	0	×	×	×	0	1
0	0	×	×	×	φ	φ
1	1	↓	0	0	no change	
1	1	↓	0	1	0	1
1	1	↓	1	0	1	0
1	1	↓	1	1	toggle	
1	1	0	×	×	no change	
1	1	1	×	×	no change	
1	1	↓	×	×	no change	

Note: "φ" means the output states are invalid, " ↓ " means the negative-edge of the clock pulse, and " × " means either low state or high state.

The D-type flip flop is a modified RS flip flop, which is illustrated in Figure 4-10-3. The characteristic equation of the D-type flip flop is expressed by

$$Q(t+1) = D \tag{4-10-2}$$

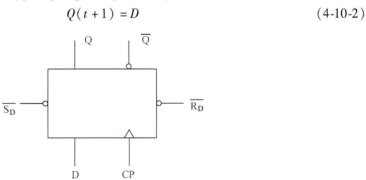

Figure 4-10-3 The logic symbol of D-type flip flop

The D-type flip flop is a positive edge trigger of clock pulse. The next output of a flip flop (or next state) can be obtained from the truth table of each type of flip flop. The truth table for the D-type flip flop is listed in Table 4-10-2.

Table 4-10-2 Truth table of positive-edge D flip flop

inputs				outputs	
set	reset	clock	D	Q	$\overline{Q}$
0	1	×	×	1	0
1	0	×	×	0	1
0	0	×	×	φ	φ
1	1	↑	1	1	0
1	1	↑	0	0	1
1	1	↑	×	no change	

Note: " ↑ " means the positive-edge of the clock pulse.

【Lab procedure】

As shown in Figure 4-10-4, there are 2 independent negative-edge-triggered JK flip flops with complementary outputs in the chip 74LS112. The five inputs and the two outputs of JK flip flop 1 are 1 $\overline{CP}$, 1K, 1J, 1 $\overline{S_D}$, and 1 $\overline{R_D}$ (pins 1, 2, 3, 4, and 15) and 1Q, 1$\overline{Q}$ (pins 5 and 6), respectively.

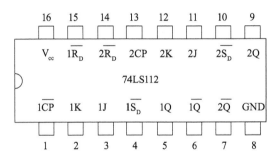

Figure 4-10-4 The pins of integrated chip 74LS112

As shown in Figure 4-10-5, there are 2 independent positive-edge-triggered D flip flops in the chip 74LS74. The four inputs and the two outputs of JK flip flop 1 are 1 $\overline{R_D}$, 1D, 1CP, 1 $\overline{S_D}$ (pins 1, 2, 3 and 4) and 1Q, 1$\overline{Q}$ (pins 5 and 6), respectively.

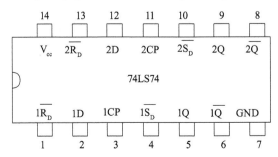

Figure 4-10-5 The pins of integrated chip 74LS74

(1) *RS flip flop*: As shown in Figure 4-10-1, connect the circuit and test "SET",

"RESET" function using the integrated chip 74LS00. Make records of the measured values, and fill them in the Table 4-10-3.

Table 4-10-3 Test RS flip flop

$\overline{S}$	$\overline{R}$	Q	$\overline{Q}$	function
0	0			
0	1			
1	0			
1	1			

(2) *JK flip flop*: As shown in Figure 4-10-2, connect the circuit and test "SET", "RESET", "J" and "K" logic function using the integrated chip 74LS112. Make records of the measured values, and fill them in Table 4-10-4.

Table 4-10-4 Test JK flip flop

pins				function							
$\overline{S_D}$	0	0	1	1							
$\overline{R_D}$	0	1	0	1							
D	×	×	×	0		0		1		1	
CP	×	×	×	↑	↓	↑	↓	↑	↓	↑	↓
Q											
$\overline{Q}$											

(3) *D flip flop*: As shown in Figure 4-10-3, connect the circuit and test "SET", "RESET", "D" and "CP" logic function using the integrated chip 74LS74. Make records of the measured values, and fill them in Table 4-10-5.

Table 4-10-5 Test D flip flop

pins				function							
$\overline{S_D}$	0	0	1	1							
$\overline{R_D}$	0	1	0	1							
J	×	×	×	0		0		1		1	
K	×	×	×	0		1		0		1	
CP	×	×	×	↑	↓	↑	↓	↑	↓	↑	↓
Q											
$\overline{Q}$											

【Pre-lab questions】

(1) Preview the concept of the flip flops.

(2) Preview the principle and function of the RS, JK and D.

(3) Preview the conversion among the RS, D and JK flip flop.

【Post-lab questions】

(1) What are the types of the flip flops?

(2) What are the roles of clock pulses of D and JK flip flop?

(3) What are the features of RS, D and JK flip flops?

(4) What is the "invalid" state of the flip flops?

4.11 Sequential logic circuits

【Lab objective】

(1) Understand the principle of a sequential logic circuit.

(2) Understand the principle of the up and down asynchronous counter.

(3) Learn to test the function of a sequential logic circuit.

【Lab devices】

Signal generator, LED, oscilloscope, regulated power supply, wires, and integrated chip 74LS112.

【Lab principle】

In contrast to the combinational circuit, the output of a sequential logic circuit depends on current external input and previous internal input binary variables. As shown in Figure 4-11-1, there are memory elements and logic gates in sequential circuits. The application fields of a sequential circuit include clocks, counters, flip-flops, memories, bistables, and registers.

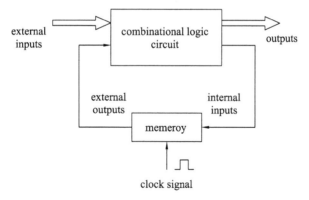

Figure 4-11-1 The block diagram of a sequential logic circuit

The design procedure of sequential logic circuits involves the following steps.

(1) Draw the state diagram from the given problem statement.

(2) Create a new reduced state table by removing all the redundant states.

(3) Determine the number of flip flops and choose the type of flip flops.

(4) Derive the output function and the flip flop input functions with map or some other simplification method.

(5) Draw a logic diagram or a list of Boolean functions.

【Lab procedure】

(1) *Up asynchronous counter*: Connect the circuit as shown in Figure 4-11-2, and switch on the power supply after checking the circuit. The Q_1, Q_2, and Q_3 outputs of each flip flop will serve as the respective binary bits of the final autputs. Observe the waveform with an oscilloscope, and draw the waveform on the graph sheet. Make records of the measured values, and fill them in Table 4-11-1.

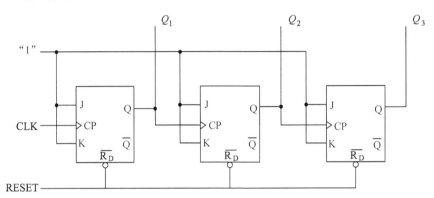

Figure 4-11-2 Three-bit binary up asynchronous counter

Table 4-11-1 Three-bit binary up asynchronous counter

RESET	CP	Q_1	Q_2	Q_3
0	0			
1	1			
1	2			
1	3			
1	4			
1	5			
1	6			
1	7			
1	8			

(2) *Down asynchronous counter*: Connect the circuit as shown in Figure 4-11-3, and switch on the power supply after checking the circuit. The Q_1, Q_2, and Q_3 outputs of each flip flop will serve as the respective binary bits of the final outputs. Observe the waveform with an oscilloscope, and draw the waveform on the graph sheet. Make records of the measured values, and fill them in Table 4-11-2.

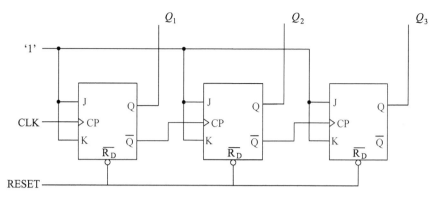

Figure 4-11-3 Three-bit binary down asynchronous counter

Table 4-11-2 Three-bit binary down asynchronous counter

RESET	CP	Q_1	Q_2	Q_3
0	0			
1	1			
1	2			
1	3			
1	4			
1	5			
1	6			
1	7			
1	8			

【Pre-lab questions】

(1) Preview the principle and function of a sequential logic circuit.

(2) Preview the principle of the up asynchronous counter.

(3) Preview the principle of the down asynchronous counter.

【Pro-lab questions】

(1) How to implement the three-digit binary up counter by D flip flops?

(2) Is the duty cycle of the clock signal allowed to be less than 50% in the asynchronous counter circuit?

(3) What is the difference between the asynchronous counter and the synchronous counter?

(4) What is the difference between the up and down asynchronous counters?

4.12 Digital counter, decoder and display circuits

【Lab objective】

(1) Understand the principle and function of counters and decoders.
(2) Understand the principle of the decoder display.
(3) Learn to use the integrated chips 74LS163 and CC4511.

【Lab devices】

Signal generator, LED, oscilloscope, regulated DC power supply, wires, integrated chips 74LS112, 74LS74 and 74LS00.

【Lab principle】

As shown in Figure 4-12-1, a binary-coded-decimal (BCD) counter is a special type of a digital counter which can count to ten on the application of a clock signal. The first stage of the counter cycle begins from the REST (0000). When a clock signal input is connected to the counter circuit, then the circuit will count the binary sequence.

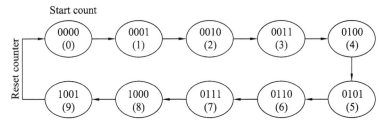

Figure 4-12-1 A BCD counter diagram

The seven segment display is used in calculators, digital counters, digital clocks, which are frequently driven by the output phases of digital integrated circuits like decade counters as well as latches. As shown in Figure 4-12-2, the decoder is an essential component in BCD to seven segments decoder. A decoder is nothing but a combinational logic circuit mainly used for converting a BCD to an equivalent decimal number. Seven segment display with CD4511 encoder is listed in Table 4-12-1.

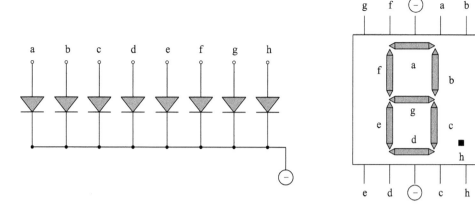

Figure 4-12-2 The seven segment display and its connected mode

Table 4-12-1 seven segment display with CD4511 encoder

input							output							
LE	$\overline{BI}$	$\overline{LT}$	D	C	B	A	a	b	c	d	e	f	g	display
×	×	0	×	×	×	×	1	1	1	1	1	1	1	8
×	0	1	×	×	×	×	0	0	0	0	0	0	0	
0	1	1	0	0	0	0	1	1	1	1	1	1	0	0
0	1	1	0	0	0	1	0	1	1	0	0	0	0	1
0	1	1	0	0	1	0	1	1	0	1	1	0	1	2
0	1	1	0	0	1	1	1	1	1	1	0	0	1	3
0	1	1	0	1	0	0	0	1	1	0	0	1	1	4
0	1	1	0	1	0	1	1	0	1	1	0	1	1	5
0	1	1	0	1	1	0	0	0	1	1	1	1	1	6
0	1	1	0	1	1	1	1	1	1	0	0	0	0	7
0	1	1	1	0	0	0	1	1	1	1	1	1	1	8
0	1	1	1	0	0	1	1	1	1	0	0	1	1	9
0	1	1	1	0	1	0	0	0	0	0	0	0	0	
0	1	1	1	0	1	1	0	0	0	0	0	0	0	
0	1	1	1	1	0	0	0	0	0	0	0	0	0	
0	1	1	1	1	0	1	0	0	0	0	0	0	0	
0	1	1	1	1	1	0	0	0	0	0	0	0	0	
0	1	1	1	1	1	1	0	0	0	0	0	0	0	
1	1	1	×	×	×	×	*	*	*	*	*	*	*	*

Note:" * " Depends upon the BCD code applied during the low to high transition of LE.

【Lab procedure】

As shown in Figure 4-12-3, the chip 74LS163 is high-speed 4-bit synchronous counters. Pin 1 is clear; pin 2 is clock pulse; pins 3 – 6 are applied to parallel inputs; pin 7 is count enable parallel input; pin 10 is count enable trickle input; pin 11 – 14 are applied to parallel outputs; pin 15 is terminal count output.

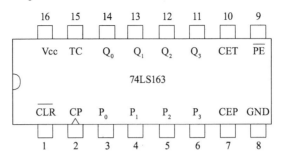

Figure 4-12-3　The diagram of the integrated chip 74LS163 pins

The CD4511 BCD-to-seven segment decoder consists of complementary MOS and NPN bipolar output drivers in a single monolithic structure. As shown in Figure 4-12-4, pins 1, 2, 6, and 7 are applied to inputs, pin 3 is used to test the display, pin 4 is used to turn-off or pulse modulate the brightness, pin 5 is used for storing BCD code, pins 9 – 15 are applied to outputs.

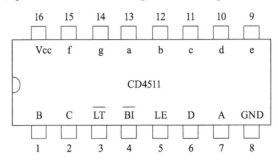

Figure 4-12-4　The diagram of integrated chip CD4511 pins

(1) *Digital counter*: As shown in Table 4-12-2, test "clear", "load", "count", "hold" and logic function using the integrated chip 74LS163. Observe the results and make records of the measured values, before the lab is over.

Table 4-12-2　The mode table of the integrated chip 74LS163

CLR	PE	CET	CEP	Function
0	×	×	×	reset (clear)
1	0	×	×	load
1	1	1	1	count
1	1	0	×	no change (hold)
1	1	×	0	no change (hold)

(2) *Decoder and display*: As shown in Figure 4-12-4, test the pins 3 and 4 logic functions using the integrated chip CD4511. Observe the results and make records of measured value.

(3) *Counter, decoder and display*: The circuit consist of the chips 74LS163 and CD4511. the output Q_0, Q_1, Q_2, Q_3 of the chip 74LS163 is connected to the inputs A, B, C, D of the chip

CD4511. Observe the results and make records of measured value.

【Pre-lab questions】

(1) Preview the principle and function of counters.

(2) Preview the principle and function of decoders.

(3) Preview the principle and function of the display.

【Pro-lab questions】

(1) How to display letters and capital letters?

(2) What are the features of the decoders?

(3) What are the features of the display?

(4) How to design decimal counters, decoders and display circuits?

4.13 555 timer applications

【Lab objective】

(1) Understand the principle and structure of the 555 timer.
(2) Understand the astable oscillator mode and monostable mode of the 555 timer.
(3) Learn to measure the pulse duration time and frequency by an oscilloscope.

【Lab devices】

Signal generator, oscilloscope, regulated DC power supply, wires, and integrated chip 555 timer.

【Lab principle】

The 555 timer is a chip that can be used to generate pulses of various duration, and to toggle between high and low states in response to inputs. By wiring with resistors and capacitors, the 555 timer is operated in astable oscillator, monostable and bistable modes. As shown in Figure 4-13-1, the 555 timer got its name from the three 5 kΩ resistors connected in a voltage divider pattern.

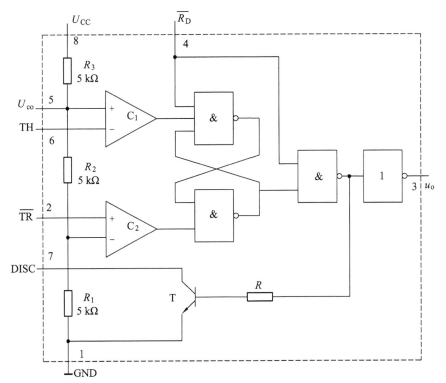

Figure 4-13-1   The inside structure of 555 timer

The circuits and waveforms of astable mode for a 555 timer are shown in Figure 4-13-2 and 4-13-3, respectively. The charge time of capacitor is expressed by

$$T_H = (R_1 + R_2) C \ln 2 \approx 0.693(R_1 + R_2) C \qquad (4\text{-}13\text{-}1)$$

The discharge time of capacitor is expressed by

$$T_L = R_2 C \ln 2 \approx 0.693 R_2 C \qquad (4\text{-}13\text{-}2)$$

The time period of output voltage is expressed by

$$T = (R_1 + 2R_2) C \ln 2 \approx 0.693(R_1 + 2R_2) C \qquad (4\text{-}13\text{-}3)$$

The duty cycle of output voltage is expressed by

$$q = \frac{R_1 + R_2}{R_1 + 2R_2} \qquad (4\text{-}13\text{-}4)$$

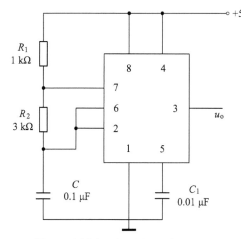
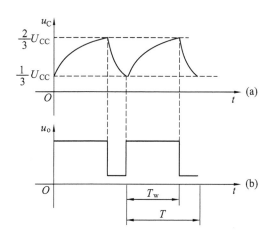

Figure 4-13-2 Astable oscillator mode   Figure 4-13-3 The capacitor voltage and output voltage

The circuits and waveforms of monostable mode for a 555 timer are shown in Figures 4-13-4 and 4-13-5, respectively. The pulse duration of the output voltage is equal to

$$T_w = RC \ln 3 \approx 1.1 RC \qquad (4\text{-}13\text{-}5)$$

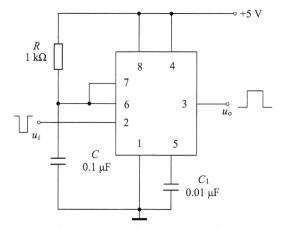

Figure 4-13-4 Monostable mode

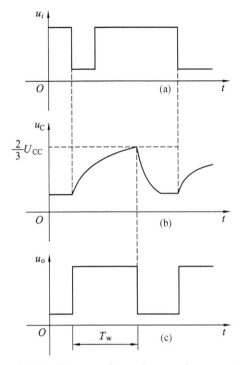

Figure 4-13-5 The capacitor voltage and output voltage

【Lab procedure】

As shown in Figure 4-13-6, carefully identify the pins of 555 timer. Firstly, carefully identify the 8 pins of the 555 timer which are listed in Table 4-13-1.

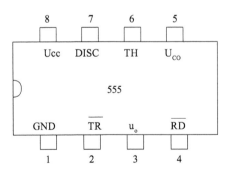

Figure 4-13-6 The 555 timer pins

Chapter 4 Electronics Lab Parts

Table 4-13-1 The pins of 555 timer

No.	name	details
1	ground	connects to the 0 v power supply
2	trigger	It works like a starter's pistol to start the 555 timer. Detects 1/3 of rail voltage to make output high.
3	output	It is either low or high (0 or 1).
4	reset	restart the 555 timer's timing operation.
5	control	It is simply connected to ground, usually through a small 0.01 μF capacitor.
6	threshold	It is to monitor the voltage across the capacitor that's discharged by pin 7.
7	discharge	It is used to discharge an external capacitor that works in conjunction with a resistor to control the timing interval.
8	supply	connects to the positive power supply

(1) *Astable oscillator mode*: Connect the circuit as shown in Figure 4-13-2, switch on the power supply after checking the circuit. Measure the parameters which are listed in Table 4-13-1. Observe the results, make records of the measured values and fill them in Table 4-13-2.

Table 4-13-2 Astable mode

items			measured value			theoretical value		
R_1	R_2	C	T	f	q	T	f	q
1 kΩ	3 kΩ	0.047 μF						
1 kΩ	3 kΩ	0.1 μF						

(2) *Mono-stable mode*: Connect the circuit as shown in Figure 4-13-4, switch on the power supply after checking the circuit. The input signal is TTL signal created from the signal generator. Measure the pulse width of the output voltage waveform. Observe the results, make records of the measured values, and draw the waveform on the graph sheet.

【Pre-lab questions】

(1) Preview the principle and structure of the 555 timer.

(2) Preview the astable oscillator mode of 555 timer.

(3) Preview the monostable mode of 555 timer.

【Pro-lab questions】

(1) What are the application fields of 555 timer?

(2) How to adjust the pulse width of the output voltage in the astable oscillator mode of a 555 timer?

(3) What is the role of the capacitor in the monostable mode of a 555 timer?

4.14 Waveform generation circuits

【Lab objective】

(1) Understand the principle of the sinusoidal wave generator.
(2) Understand the principles of square wave generator and triangle wave generator.
(3) Learn to design sinusoidal, square and triangle wave using op amps.

【Lab devices】

Resistors, capacitors, oscilloscope, regulated DC power supply, op amps, and wires.

【Lab principle】

An oscillator is a circuit that generates a specific periodic waveform, such as square wave, triangular wave, sawtooth wave, and sine wave. An oscillator generates output without any AC input signal. An electronic oscillator is a circuit which converts DC energy into AC at a very high frequency. Sinusoidal oscillators can be classified into tuned circuit oscillators, RC oscillators, crystal oscillators and negative-resistance oscillators. The basic structure of RC sinusoidal oscillator consists of RC series-parallel circuits and an, op amps, which is illustrated in Figure 4-14-1.

Negative feedback network consists of R_3, R_4 and a variable resistor R_w, which can adjust the voltage gain of amplifier circuits. The role of the diodes D_1 and D_2 are to stabilize the amplitude of the output voltage. Since the resistor R_2 and the capacitor C_2 are connected in series and the resistor R_1 and the capacitor C_1 are connected in parallel (a limiting condition: $R_2 = R_1$, $C_2 = C_1$), it has a frequency selection characteristic. Oscillating frequency is expressed by

$$f_0 = \frac{1}{R_1 C_1} \qquad (4\text{-}14\text{-}1)$$

The circuit of square wave and triangle wave generator is illustrated in Figure 4-14-2. A square wave is created by a hysteresis comparator. A triangle wave can be achieved by the integration of a square wave. Therefore, RC integrator is implemented by an op amp.

Chapter 4 Electronics Lab Parts

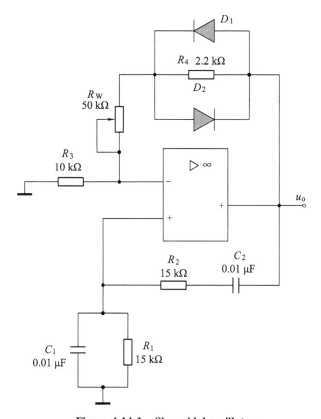

Figure 4-14-1 Sinusoidal oscillators

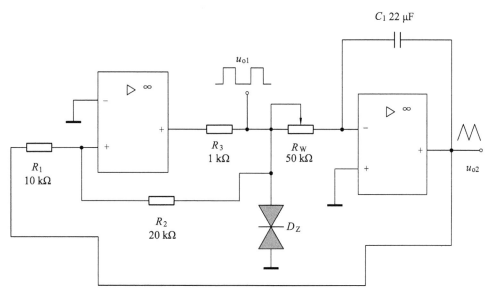

Figure 4-14-2 Square wave and triangle wave generator

【Lab procedure】

(1) *Sinusoidal wave*: Connect the circuit as shown in Figure 4-14-1, switch on the power

supply after checking the circuit. Adjust the variable resistor R_w, observe the results, make records of the measured values, and draw the curve on the graph sheet.

(2) *Square wave and triangle wave*: Connect the circuit as shown in Figure 4-14-2, switch on the AC power supply after checking the circuit. Adjust the variable resistor R_w, observe the results, make records of the measured values, draw the curve on the graph sheet.

【Pre-lab questions】

(1) Preview the principle of the sinusoidal wave generator.

(2) Preview the principle of square wave generator.

(3) Preview the principle of triangle wave generator.

【Pro-lab questions】

(1) How to adjust the frequency of the sinusoidal wave?

(2) How to adjust the period of the square wave?

(3) What is the role of the capacitor in the triangle wave generator?

Chapter 5

NI ELVISmx Introduction

National instrument (NI) offers a combination of productive software, high-quality drivers, and modular hardware in automated test and automated measurement systems. Teaching-oriented products mainly include ELVIS workstation, myDAQ device, and LabVIEW software. A myDAQ device is a data acquisition module that features commonly used plug-and-play lab instruments. It is mainly used for project-based learning and hands-on experimentation for analog circuits, sensors, signals, and systems. LabVIEW is a kind of graphical programming language that uses icons instead of lines of text to create applications, including front panel window and block diagram window.

The national instrument educational laboratory virtual instrumentation suite (NI ELVIS) is a modular engineering educational lab device developed for academia, such as electrical engineering, mechanical engineering, and biomedical engineering. The NI ELVIS is suitable for teaching basic electronics and circuit design, and provides the complete test, measurement and data logging capabilities. As well as soft front panels on the NI ELVIS instrument launcher strip, NI ELVIS can use LabVIEW and MATLATB to assist the measurement. In other words, ELVIS introduces students to engineering problem-solving with portable and programmable measurements hands-on experimentation in or outside the lab. With mobile NI ELVIS workstations, lab space is more efficient and the instrument cost is also reduced. It is benefit to the management of electrical and electronics lab.

5.1 NI ELVISmx workstation

5.1.1 Workstation connectors and controls

A typical of NI ELVISmx series system consists of laptop computer, USB cable, and NI

ELVISmx workstation. The NI ELVISmx workstation includes prototyping board, DMM fuse and connectors, oscilloscope connectors, function generator output and digital trigger input connector, prototyping board mounting screw holes, prototyping board connector, prototyping board power switch, status LEDs, variable power supplies manual controls, and function generator manual controls.

USB LEDs have two state, which is listed in Table 5-1-1. USB LEDs "ready" means that the NI ELVISmx series hardware is properly configured and ready to communicate with the host computer. USB LEDs "active" indicates activity on the USB connection to the host computer.

Table 5-1-1 USB LED modes

active LED	ready LED	description
off	off	main power is off.
yellow	off	no connection to the host computer is detected.
off	green	connected to a full speed USB host.
off	yellow	connected to a high-speed USB host.
green	green or yellow	communicate with host.

Variable power supplies controls have positive voltage adjust knob and the negative voltage adjust knob. Positive voltage adjust knob is to control the output voltage of the positive variable power supply. The range of the positive supply is between 0 to +12 V. Similarly, the negative voltage adjust knob is to control the output voltage of the negative variable power supply. The range of the negative supply is betwwen 0 and −12 V. These knobs are effective only when the associated variable power supply is set to manual mode. When the variable power supply is in manual mode, the LED next to each knob lights up.

Function generator controls have frequency knob and amplitude knob. Frequency knob is used to adjust the output frequency of the generated waveform. Amplitude knob is used to adjust the amplitude of the generated waveform. Note these knobs are only active when the function generator is set to manual mode. A LED between the knob lights when the function generator is in manual mode.

Oscilloscope connectors have CH0 and CH1 BNC connector. CH0 BNC connector is the input for channel 0 of the oscilloscope. Similarly, CH1 BNC connector is the input for channel 1 of the oscilloscope. Note the NI ELVISmx oscilloscope channels 0 and 1 are available only through the BNC connectors. They are not internally routed to the prototyping board.

DMM connectors have red banana jack and common banana jack. Banana jack (red) is the positive input of voltage, current, resistance, and diode. Common banana jack (black) is the common reference or the negative input. In addition, voltage, current, resistance, and diode measurements are available only through the banana jacks (the prototyping board isn't allowed).

Finally, the workstation rear panel includes the workstation power switch, an AC/DC power supply connection, a USB port, a cable tie slot and a Kensington security cable lock connector.

5.1.2　Prototyping board

Using prototyping boards, NI ELVISmx workstations can be interchanged. The prototype boards provide areas for building electronic circuits. The prototype board can be connected to NI ELVISmx signal terminals through the power distribution boards on both sides of the breadboard area. Also, be sure to turn off the prototype board power switch before plugging it into or removing it from the workstation. The NI ELVISmx prototyping board has four solderless breadboards integrated to rails on either side of the breadboard space, which includes sockets, bus strip, terminal strips and center divider.

(1) *Sockets*: They are holes in the board with a spacing of 0.1 cm between the two closest holes. Inside each socket, there is a metal clip that secures the component lead when you insert it into the hole.

(2) *Terminal strips*: The area of building circuit is referred to as a terminal strip. There are sixty-four rows divided into 10 individual strips named A, B, C, D, E and F, G, H, I, J. These rows are separated by a center divider. For each row, the strips A, B, C, D, E are electronically connected to a node. Similarly, F, G, H, I, J are connected into a singular node. It is allowed to appropriately connect wires and devices without having to use one socket for multiple items.

(3) *Bus strip*: There are four bus strips, two on each side with the label " + " and " − ". The bus strips are also commonly referred to as power rails because power and ground signals are usually connected to it.

(4) *Center divider*: The spacing between the socket E and F is the center divider. It is designed with enough spacing for you to place a dual-line integrated circuit that would connect across the center divider and connect into the appropriate strips (A-E or F-J).

With a prototyping board, the steps to build the circuit are briefly given as follows:

(1) Use the bus or power strips to connect input signals, output paths or ground.

(2) Use the terminal strips to connect circuit topology, such as resistors, capacitors, diodes, transistors, and inductors.

(3) Use the center divider to place an integrated circuit, such as an operational amplifier.

(4) Connect instruments to the circuit and make measurements.

5.2　NI ELVISmx instrument launcher

The NI ELVISmx instrument launcher provides access to the instruments, documentation and online resource links. Start the instrument launcher by navigating to *Start→All Program Files→ National Instruments→NI ELVISmx→NI ELVISmx Instrument Launcher*.

As shown in Figure 5-2-1. The NIELVIS instrument launcher includes arbitrary waveform generator (ARB), bode analyzer, digital reader, digital writer, digital multimeter (DMM),

The Lab Tutorial of Electrical Engineering and Electronics

dynamic signal analyzer (DSA), function generator (FGEN), oscilloscope, 2-wire current-voltage analyzer, 3-wire current-voltage analyzer, impedance analyzer, variable power supplies, digital waveform viewer, 8-channel oscilloscope, audio equalizer, data logger, DC level, and octave analyzer.

Figure 5-2-1 NI ELVISmx instrument launcher

To start an instrument, click the button corresponding to the desired instrument. If two instruments with overlapping functions are running simultaneously, NI ELVISmx software will generate an error dialog. The instrument will be disabled, and does not work until conflict is solved. NI ELVISmx resource conflicts is listed in Table D-1 (Appendix D).

(1) *Arbitrary waveform generator*: The soft front panel of arbitrary waveform generator is illustrated in Figure 5-2-2. This advanced instrument uses the analog output (AO) capabilities of

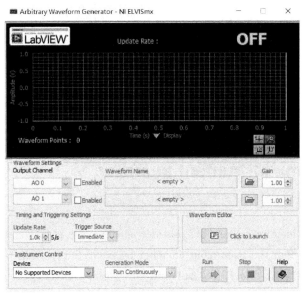

Figure 5-2-2 The soft front panel of arbitrary waveform generator

the device. You can use the waveform editor software to create various signal types and load them into ARB. Since the device has two AO channels, two waveforms can be generated simultaneously. It can run continuously or once.

(2) *Bode analyzer*: The soft front panel of bode analyzer is illustrated in Figure 5-2-3. A Bode analyzer describes the frequency response of a circuit by showing the gain and phase as a function of frequency. Using a function generator in conjunction with analog input (AI), the NI ELVISmx Bode analyzer can calculate the frequency response of a circuit. In addition, linear and logarithmic scales of the gain graph are provided.

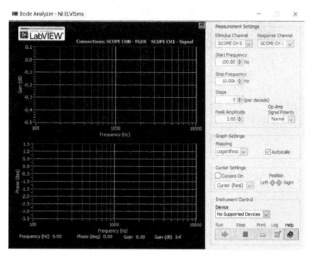

Figure 5-2-3 **The soft front panel of bode analyzer**

(3) *Digital reader*: The soft front panel of digital reader is illustrated in Figure 5-2-4. This instrument reads digital data from the NI ELVIS series digital lines. You can read eight consecutive lines at a time: {0, 1, 2, 3, 4, 5, 6, 7}; {8, 9, 10, 11, 12, 13, 14, 15}; {16, 17, 18, 19, 20, 21, 22, 23} either continuously or you can take a single reading.

Figure 5-2-4 **The soft front panel of digital reader**

(4) *Digital writer*: The soft front panel of digital writer is illustrated in Figure 5-2-5. This instrument updates the digital lines with user-specified digital patterns. You can manually create a pattern or select predefined modes, such as ramp, toggle, or walking. This instrument can control eight consecutive lines and either continually output a pattern or just perform a single write. Output voltage levels are TTL compatible.

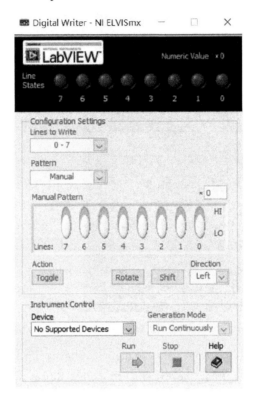

Figure 5-2-5　The soft front panel of digital writer

(5) *Digital multimeter*: The soft front panel of digital multimeter is illustrated in Figure 5-2-6. The top row of nine buttons denotes the different DMM modes, namely from left to right: DC voltage measurement, AC voltage measurement, DC current measurement, AC current measurement, resistance measurement, capacitance measurement, inductance measurement, diode continuity and audible continuity. If the capacitor and inductor are measured, capacitor and inductor should be connected to the DMM on the prototyping board. If the other electronics parameters are measured, the corresponding elements should be connected to the DMM banana jacks.

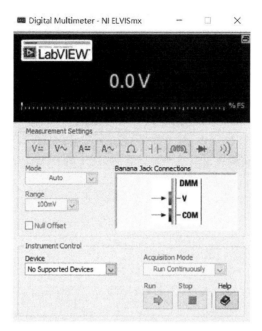

Figure 5-2-6 The soft front panel of digital multimeter

(6) *Dynamic signal analyzer*: The soft front panel of dynamic signal analyzer is illustrated in Figure 5-2-7. This instrument performs a frequency domain transform of the AI or scope waveform measurement. It can either continuously make measurements or scan one time. In addition, dynamic signal analyzer provides to windowing and averaging options.

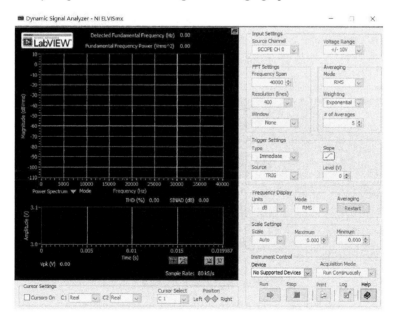

Figure 5-2-7 The soft front panel of dynamic signal analyzer

(7) *Function generator*: The soft front panel of function generator is illustrated in Figure 5-2-8. This instrument generates standard waveforms and provides options for output waveform

(sinusoidal, rectangle, and triangle), with amplitude selection and frequency settings. The function generator offers DC offset setting, frequency sweep capabilities, amplitude modulation, and frequency modulation. In addition, there are the FGEN to the prototyping board or to the FGEN and TRIG BNC connector on the left side of the workstation.

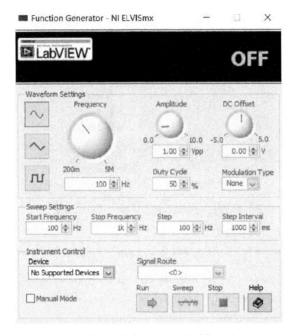

Figure 5-2-8　The soft front panel of function generator

(8) *Impedance analyzer*: The interface of impedance analyzer is illustrated in Figure 5-2-9. This instrument is capable of measuring the resistance and reactance of passive 2-wire components at a given frequency. The device is commonly used to explain sinusoidal steady state analysis.

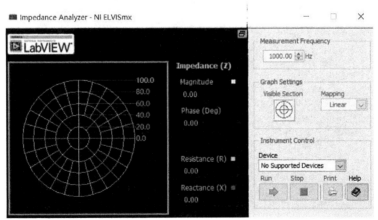

Figure 5-2-9　The soft front panel of impedance analyzer

(9) *Oscilloscope*: The soft front panel of the oscilloscope is illustrated in Figure 5-2-10. The oscilloscope has two channels and provides scaling, position adjustable knob and adjustable time base. Oscilloscope provides the trigger source, digital or analog hardware triggering, and mode

settings. With auto zoom function, the voltage display ratio based on the peak-to-peak voltage can be adjusted. In addition, the oscilloscope from the BNC connectors on the left side panel of the workstation can be used.

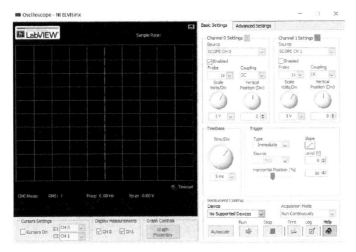

Figure 5-2-10　The soft front panel of the oscilloscope

(10) *Two-wire current-voltage analyzers*: The soft front panel of the two-wire current-voltage analyzers is illustrated in Figure 5-2-11. The two-wire voltage analyzer is used to conduct parametric testing of diodes in the form of current-voltage curves. A common use of this device is to determine the forward bias voltage and small signal resistance for the piece-wise approximation of a test diode. Two-wire meter provides full flexibility in setting parameters such as voltage and current ranges and can save the data to a file. In addition, the instrument has cursors in order to implement more accurate screen measurement.

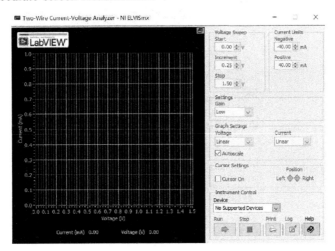

Figure 5-2-11　The soft front panel of the two-wire current-voltage analyzer

(11) *Three-wire current-voltage analyzers*: The soft front panel of the three-wire current-voltage analyzers is illustrated in Figure 5-2-12. The three-wire voltage analyzer is used to conduct parametric testing of transistors in the form of characteristic curves. Three-wire meters

provide the base current settings for NPN and PNP transistor measurements. In addition, the instrument has cursors in order to implement more accurate screen measurement.

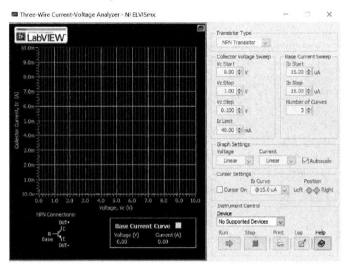

Figure 5-2-12   The soft front panel of the three-wire current-voltage analyzer

(12) *Variable power supplies*: The soft front panel of the variable power supplies is illustrated in Figure 5-2-13. This instrument can offer the output of the positive or negative variable power supply. The range of the negative power supply is between −12 and 0 V, and the range of the positive pow supply is between 0 and +12 V.

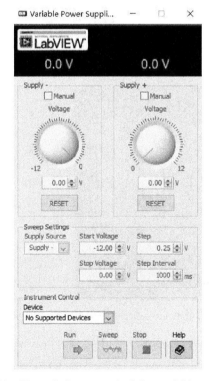

Figure 5-2-13   The soft front panel of the variable power supplies

5.3 Interface with MATLAB

5.3.1 NI ELVISmx with data acquisition toolbox

In the electrical engineering and electronics academic community, MATLAB are commonly used for numeric analysis and visualization, includes signal processing toolbox, data acquisition (DAQ) toolbox, instrument control toolbox, control system toolbox. Data acquisition toolbox provides the functions for configuring data acquisition hardware, reading data into MATLAB and simulink, and writing data to DAQ analog and digital output channels. The toolbox supports a variety of data acquisition hardware, including USB, peripheral component interface, PXI, and PXI express devices from national instruments and other manufactures.

Data acquisition toolbox provides the following library functions for interaction between MATLAB and NI ELVISmx prototyping platform.

  elvis. setup
  elvis. Fgen
  elvis. FgenBNC
  elvis. Multimeter
  elvis. MultimeterAuto
  elvis. private. DMMCurrentAC
  elvis. VarPowSupply
  elvis. private. DMMCurrentDC
  elvis. private. DMMVoltageDC
  elvis. private. NIDAQmx_hdr
  elvis. private. Instrument
  elvis. private. Const
  elvis. private. TestNIDAQmx
  elvis. private. DMMSuper
  elvis. util. checkClasses
  elvis. private. NIDAQmx
  elvis. private. DMMVoltageAC
  elvis. private. DMMResistance
  elvis. private. DMMContinuity
  elvis. private. DMMDiode
  elvis. util. checkClasses
  elvis. util. getErrorInfo
  elvis. util. getErrorString

With library functions in the data acquisition toolbox, you can configure the parameters of the variable power supply of NI ELVISmx platform. The codes are given as follows.

% ------ Set variable power supply ------
vps = elvis. VarPowSupply;
vps. Vpos = 7;
vps. Vneg = -5;
delete(vps);

To configure the parameters of the function generator, the codes are given as follows.

% ------ Set function generator ------
fg = elvis. Fgen;
fg. Function = 'triangle';
fg. Amplitude = 3;
fg. Frequency = 200;
fg. start( );
pause(1);
fg. stop( );
delete(fg);

To configure the parameters of the digital multimeter, the codes are given as follows.

% ------ Set digital multimeter ------
dmm = elvis. Multimeter('dcvoltage');
dmm. Range = '10V';
data = dmm. readData( );
delete(dmm);

5.3.2 Lab: control stepper motor using digital outputs

The hardware system consists of a Portescap 20M020D1U motor, a DB25 male solder connector, and a Texas instruments ULN2003 digital outputs. The block diagram is illustrated in Figure 5-3-1. The DB25 male solder connectors is a kind of terminated connection cable, which can easily be soldered to a custom length cable and be created a custom pin out.

The parameters of 20M020D1U motor is given as follows: rated voltage of a Portescap 20M020D1U is 5 V; resistance is 20 Ω; inductance per phase is 3.9 mH; rated current per phase is 0.25 amps; step angle is 18 degree; part number is unipolar. The ULN2003 devices include 7 NPN Darlington transistor arrays, which can be paralleled for higher current capability. As as result, it is characterized by high-voltage and high-current outputs with common-cathode clamp diodes for switching inductive loads. The collector-current rating of a single Darlington pair is 500 mA. The application fields include relay drivers, LED display drivers, line drivers, and logic buffers.

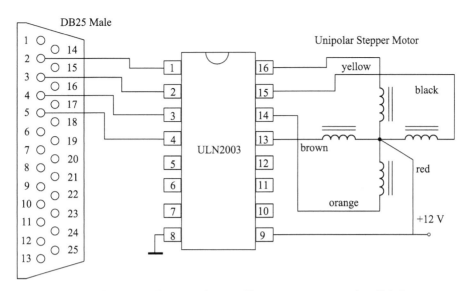

Figure 5-3-1 The block diagram of controlling stepper motor using digital outputs

To control stepper motor using digital outputs, the MATLAB code is as follows:

(1) *To discover a device that supports digital output*:

devices = daq. getDevices

devices =

Data acquisition devices:

index	Vendor	Device ID	Description
1	ni	cDAQ1Mod1	National Instruments NI 9205
2	ni	cDAQ1Mod2	National Instruments NI 9263
3	ni	cDAQ1Mod3	National Instruments NI 9234
4	ni	cDAQ1Mod4	National Instruments NI 9201
5	ni	cDAQ1Mod5	National Instruments NI 9402
6	ni	cDAQ1Mod6	National Instruments NI 9213
7	ni	cDAQ1Mod7	National Instruments NI 9219
8	ni	cDAQ1Mod8	National Instruments NI 9265
9	ni	Dev1	National Instruments PCIe-6363
10	ni	Dev2	National Instruments NI ELVIS

devices (10)

ans =

ni: National Instruments NI ELVIS II (Device ID: 'Dev2')

Analog input subsystem supports:
   7 ranges supported
   Rates from 0.0 to 1250000.0 scans/sec
   16 channels ('ai0' - 'ai15')
   'Voltage' measurement type

Analog output subsystem supports:
   -5.0 to +5.0 Volts, -10 to +10 Volts ranges
   Rates from 0.0 to 2857142.9 scans/sec
   2 channels ('ao0','ao1')
   'Voltage' measurement type
Digital subsystem supports:
   39 channels ('port0/line0'-'port2/line6')
   'InputOnly','OutputOnly','Bidirectional' measurement types

Counter input subsystem supports:
   Rates from 0.1 to 80000000.0 scans/sec
   2 channels ('ctr0','ctr1')
   'EdgeCount' measurement type

Counter output subsystem supports:
   Rates from 0.1 to 80000000.0 scans/sec
   2 channels ('ctr0','ctr1')
   'PulseGeneration' measurement type

(2) *Create a session, and add 4 digital channels on port* 0, *lines* 2 - 5: Set the measurement type to OutputOnly. These are connected to the four control lines for the stepper motor.

s = daq.createSession('ni');
addDigitalChannel(s,'Dev2','port0/line2:5','OutputOnly')
ans =

1 × 604 char array

Data acquisition session using National Instruments hardware: clocked operations using startForeground and startBackground are disabled. Only on-demand operations using inputSingleScan and outputSingleScan can be done.
   Number of channels: 4
   index Type Device   Channel   MeasurementType   Range name
   ----
   1   dio   Dev2   port0/line2   OutputOnly   n/a

2	dio	Dev2	port0/line3	OutputOnly	n/a
3	dio	Dev2	port0/line4	OutputOnly	n/a
4	dio	Dev2	port0/line5	OutputOnly	n/a

(3) *Define motor steps*: Refer to the Portescap motor wiring diagram describing the sequence of 4 bit patterns. Send this pattern sequentially to the motor to produce counterclockwise motion. Each step turns the motor 18 degrees. Each cycle of 4 steps turns the motor 72 degrees. Repeat this sequence five times to rotate the motor 360 degrees.

step1 = [1 0 1 0];
step2 = [1 0 0 1];
step3 = [0 1 0 1];
step4 = [0 1 1 0];

(4) *Rotate motor*: use outputSingleScan to output the sequence to turn the motor 72 degrees counterclockwise.

outputSingleScan(s, step1);
outputSingleScan(s, step2);
outputSingleScan(s, step3);
outputSingleScan(s, step4);

(5) *Repeat sequence* 50 *times to rotate the motor* 10 *times counterclockwise*:

for motorstep = 1:50
    outputSingleScan(s, step1);
    outputSingleScan(s, step2);
    outputSingleScan(s, step3);
    outputSingleScan(s, step4);
end

(6) *To turn the motor 72 degrees clockwise, reverse the order of the steps*:

outputSingleScan(s, step4);
outputSingleScan(s, step3);
outputSingleScan(s, step2);
outputSingleScan(s, step1);

(7) *Turn off all outputs*: after using use the motor, turn off all the lines to allow the motor to rotate freely.

outputSingleScan(s, [0 0 0 0]);

Appendix

Appendix A Electrical and electronic units

Table A-1 Electrical and electronic units

unit name	unit symbol	description
ampere	A	electric current
volt	V	voltage
ohm	Ω	resistance
watt	W	electric power
decibel-milliwatt	dBm	electric power
decibel-watt	dBW	electric power
volt-ampere-reactive	var	reactive power
volt-ampere	VA	apparent power
farad	F	capacitance
henry	H	inductance
siements/mho	S	conductance/admittance
coulomb	C	electric charge
joule	J	energy
kilowatt-hour	kWh	energy
ohm-meter	$\Omega \cdot m$	resistivity
siemens per meter	S/m	conductivity
volts per meter	V/m	electric field
volt-meter	$V \cdot m$	electric flux
tesla	T	magnetic field
weber	Wb	magnetic flux
decibel	dB	a ratio

Appendix B Metric prefix

Table B-1 Metric prefix

metric prefix	symbol symbol	exponential	description
Yotta	Y	10^{24}	septillion
Zetta	Z	10^{21}	sextillion
Exa	E	10^{18}	quintillion
Peta	P	10^{15}	quadrillion
Tera	T	10^{12}	trillion
Giga	G	10^{9}	billion
Mega	g M	10^{6}	million
kilo	k	10^{3}	thousand
hecto	h	10^{2}	hundred
deca	da	10^{1}	ten
base	b	10^{0}	one
deci	d	10^{-1}	tenth
centi	c	10^{-2}	hundredth
milli	m	10^{-3}	thousandth
micro	μ	10^{-6}	millionth
nano	n	10^{-9}	billionth
pico	p	10^{-12}	trillionth
femto	f	10^{-15}	quadrillionth
atto	a	10^{-18}	quintillionth
zepto	z	10^{-21}	sextillionth
yocto	y	10^{-24}	septillionth

Appendix C Abbreviations in electrical engineering and electronics

Table C-1 Abbreviations in electrical engineering and electronics

full words	symbol	full words	symbol
amperes, amps	A	number of turns in a winding	N
alternating current	AC	open, off or stop	O
air circuit breaker	ACB	over frequency	OF
audio frequency	AF	overload	OL
bandwidth	BW	over voltage	OV
electrical capacitance	C	programmable controller	PC
circuit breaker	CB	pulse amplitude modulation	PAM
direct current	DC	power factor	PF
earth or ground	E/g	phases of an electrical circuit	ph
electromagnetic compatibility	EMC	programmable logic controller	PLC
electromagnetic interference	EMI	pulse width modulation	PWM
electromotive force	EMF	reactive power	Q
electrical conductance	G	quantity	QTY
electrical reactance	X	required	REQ
high voltage, above 600 volts	HV	root mean squared	RMS
current in amperes	I	electrical resistance	R
input or output signals	I/O	radio frequency	RF
energy in joules	J	apparent power	S
electrical inductance	L	short circuit	SC
light emitting diode	LED	single phase	SP
low frequency	LF	switch	SW
low voltage, 51 to 599 volts	LV	transistor transistor logic	TTL
motor control unit	MCU	voltage	U/V
magneto-motive force	MMF	ultra-high frequency	UHF
medium voltage	MV	active power	W
normally closed	NC	electrical admittance	Y
normally open	NO	electrical impedance	Z

Appendix D NI ELVIS resource conflicts

Table D-1 NI ELVIS resource conflicts

Units of NI ELVIS platform	DMM-voltmeter, ammeter, resistance, continuity, diode	DMM-inductance, capacitance	oscilloscope (NI ELVIS and AI channels)	oscilloscope (NI ELVIS II + high speed channels)	oscilloscope digital trigger input BNC	function generator prototyping board	function generator BNC	function generator manual mode	variable power supply software mode	variable power supply manual mode	bode analyzer	dynamic signal analyzer	arbitrary waveform generator AO 0	arbitrary waveform generator AO 1	impedance analyzer	two-wire current-voltage analyzer	three-wire current-voltage analyzer
DMM-voltmeter/ammeter/resistance/continuity/diode	-	×	-	-	-	-	-	-	-	-	-	-	-	-	-	-	-
DMM-inductance, capacitance	×	-	×	-	-	×	×	-	-	×	×	×	×	-	×	×	×
oscilloscope (NI ELVIS and AI channels)	-	×	-	×	-	-	-	-	-	×	×	×	-	-	×	×	×
oscilloscope (NI ELVIS II + high speed channels)	-	-	×	-	-	-	-	-	-	-	-	-	-	-	-	-	-
oscilloscope digital trigger input BNC	-	-	-	-	-	-	×	-	-	-	-	-	-	-	-	-	-
function generator prototyping board	-	×	-	-	-	-	×	-	-	×	-	-	×	-	×	×	×
function generator BNC	-	×	-	-	×	×	-	-	-	×	-	×	-	-	×	×	×
function generator manual mode	-	-	-	-	-	-	-	-	-	-	-	-	-	-	-	-	-
variable power supply software mode	-	-	-	-	-	-	-	-	-	-	-	-	-	-	-	-	-
variable power supply manual mode	-	#	#	-	-	-	-	-	-	-	#	#	-	-	#	#	#
bode analyzer	-	×	×	-	-	×	×	-	-	#	-	×	-	-	×	×	×
dynamic signal analyzer	-	×	×	-	-	-	-	-	-	#	-	-	-	-	-	×	×
arbitrary waveform generator AO 0	-	×	-	-	-	×	-	-	-	×	-	-	-	-	×	×	×
arbitrary waveform generator AO 1	-	-	-	-	-	-	-	-	-	-	-	-	-	-	-	-	-
impedance analyzer	-	×	×	-	-	×	×	-	-	#	×	×	×	-	-	×	×
two-wire current-voltage analyzer	-	×	×	-	-	×	×	-	-	#	×	×	×	-	×	-	×
three-wire current-voltage analyzer	-	×	×	-	-	×	×	-	-	#	×	×	×	-	×	×	-

Note: " - " no conflicts; " × " conflict exist; "#" conflict exists if measure actual voltages option is enabled in VPS manual mode.

Appendix E NI ELVIS signal description

Table E-1 NI ELVIS signal description

signal name	type	description
AI $<0\cdots7>\pm$	analog inputs	analog inputs channels 0 through 7; "$\pm$" positive and negative input channels
AI SENSE	analog inputs	analog input sense—reference for the analog channels in NRSE mode
AI GND	analog inputs	analog input ground
BASE	3-Wire voltage	base excitation for bipolar junction transistors
DUT +	DMM, impedance, 2/3-Wire analyzers	excitation terminal for capacitance and inductance
DUT " $-$ "	DMM, impedance, 2/3-Wire analyzers	excitation terminal for capacitance and inductance
AO $<0..1>$	analog outputs	analog output channels 0 and 1 used for the arbitrary waveform generator
FGEN	function generator	function generator output
SYNC	function generator	TTL output synchronized to the FGEN signal.
BANANA $<A..D>$	user configurable I/O	banana jacks A through D
BNC $<1..2>\pm$	user configurable I/O	BNC connectors 1 and 2
SUPPLY +	variable power supplies	positive variable power supplies; output of 0 to 12 V
SUPPLY-	variable power supplies	negative variable power supplies; output of -12 V to 0
DIO $<0\cdots23>$	digital input/output	digital lines 0 through 23
+15 V	DC power supplies	+15 V fixed power supply
−15 V	DC power supplies	−15 V fixed power supply
+5V	DC Power Supplies	+5V Fixed Power Supply.
AM	function generator	amplitude modulation input—analog input used to modulate the amplitude of the FGEN signal.
FM	function generator	frequency modulation input—analog input used to modulate the frequency of the FGEN signal.
LED $<0..7>$	user configurable I/O	LEDs 0 through 7—Apply 5 V for 10 mA device
PFI8/CTR0_SOURCE	programmable function interface	Static Digital I/O, line P2.0 PFI8, Default function: Counter 0 Source
PFI9/CTR0_GATE	programmable function interface	Static Digital I/O, line P2.1 PFI9, Default function: Counter 0 Gate

Appendix

continued

signal name	type	description
PFI12/CTR0_OUT	programmable interface function	Static Digital I/O, line P2.4 PFI12, Default function: Counter 0 Out
PFI3/CTR1_SOURCE	programmable interface function	Static Digital I/O, line P1.3 PFI3, Default function: Counter 1 Source
PFI4/CTR1_GATE	programmable interface function	Static Digital I/O, line P1.4 PFI4, Default function: Counter 1 Gate
PFI13/CTR1_OUT	programmable interface function	Static Digital I/O, line P2.5 PFI13, Default function: Counter 1 Out
PFI14/FREQ_OUT	programmable interface function	Static Digital I/O, line P2.6 PFI14, Default function: Frequency Output
PFI8/CTR0_SOURCE	Programmable Interface Function	Static Digital I/O, line P2.0 PFI8, Default function: Counter 0 Source
DSUB SHIELD	user configurable I/O	Connection to D-SUB shield.
DSUB PIN <1..9>	user configurable I/O	Connections to D-SUB pins.
PFI <0..2>, <5..7>, <10..11>	programmable interface functions	PFI Lines—used for static DIO or for routing timing signals.

Reference

[1] Li Yanling, Feng Yu, Ma Qiuming. Lab Tutorial of Electrical and Electronic Technology. Chinese Edition. Beijing: Tsinghua University Press, 2017.

[2] Wang Qingchun, Liang Junlong, Chen Shouman. Lab Tutorial of Electrical and Electronic Technology. Chinese Edition. Beijing: China Science Press, 2019.

[3] Zhang Yueqin, Guo Minli. Lab Tutorial of Electrical and Electronic Technology. Chinese Edition. Chengdu: Southwest Jiaotong University Press, 2016.

[4] Qin Zenghuang, Jiang Sanyong. Electrical Engineering. 7th Edition. Chinese edition. Beijing: Higher Education Press, 2009.

[5] Zhao Buhui, Jing Liang. Electrical Engineering Ⅱ Electronic Technology. Chinese Edition. Zhenjiang: Jiangsu University Press, 2016.

[6] Allan R H. Principles and Applications of Electrical Engineering. 7th Edition. Beijing: Electronic Industry Press, 2019.

[7] Zheng Minghui, Hu Ying. The Experiment of Electrical and Electronic Technology. 2nd Edition(Chinese Edition). Beijing: Posts & Telecom Press, 2015.

[8] Tong Shibai, Hua Chengying. Analog Electronic Technology Foundation. 5th Edition. (Chinese Edition). Beijing: Higher Education Press, 2015.

[9] Wang Zhen. Lab Tutorial of Analog Electronic Technology. Chinese Edition. Beijing: China Machine Press, 2018.

[10] Yan Shi. Digital Electronic Technology Foundation. 5th Edition (Chinese Edition). Beijing: Higher Education Press, 2011.

[11] Guo Hong. Digital Electronics Technology and Application. Chinese Edition. Beijing: Posts & Telecom Press, 2019.

[12] Bai Xuefeng, Lu Xusheng. Lab Tutorial of Electrical and Electronic Technology. Chinese Edition. Beijing: China Electric Power Press, 2018.

[13] Huang Xiaoqing. Signal and System Tutorial Based on NI ELVIS. Beijing: China Machine Press, 2016.

[14] Wang Xiuping, Yu Jinhua, Lin Lili. LabVIEW and NI-ELVIS Lab Tutorial. Hangzhou: Zhejiang University Press, 2012.